JOEL B. PREDD, JON SCHMID, ELIZABETH M. BARTELS,
JEFFREY A. DREZNER, BRADLEY WILSON, ANNA JEAN WIRTH,
LIAM MCLANE

Acquiring a Mosaic Force

Issues, Options, and Trade-Offs

Prepared for the Office of the Secretary of Defense
Approved for public release; distribution unlimited

NATIONAL DEFENSE RESEARCH INSTITUTE

For more information on this publication, visit www.rand.org/t/RRA458-3

Library of Congress Cataloging-in-Publication Data is available for this publication.
ISBN: 978-1-9774-0698-9

Support RAND
Make a tax-deductible charitable contribution at
www.rand.org/giving/contribute

www.rand.org

About This Report

The Defense Advanced Research Projects Agency (DARPA) has an ambitious vision for *Mosaic Warfare*, conceived by the Strategic Technology Office leadership as both a warfighting concept and a means to greatly accelerate capability development and fielding. Mosaic Warfare entails a more fractionated, more heterogeneous force that can be dynamically composed on tactical timelines into unique force packages to surprise and overwhelm an adversary. As a result, Mosaic Warfare entails shifting away from a focus on monolithic platforms, which are slow to develop and slow to field, to a focus on simpler force elements that can be developed and fielded quickly and integrated at mission execution. Although the success of Mosaic Warfare depends on DARPA advancing multiple technologies, the Mosaic vision is inherently more challenging to transition than is a program or technology.

Anticipating this challenge, DARPA asked that the RAND Corporation examine the opportunities and challenges associated with developing and fielding a Mosaic force under existing or alternative governance models and management processes, as would be required for the vision to move from DARPA to widespread acceptance by the U.S. Department of Defense. To this end, the RAND research team designed and executed a policy game that immersed DARPA representatives and RAND analysts in the task of fielding a Mosaic force and required them to operate within the authorities, responsibilities, and constraints provided under the existing governance model and an alternative model.

The research reported here was completed in April 2021 and underwent security review with the sponsor and the Defense Office of Prepublication and Security Review before public release.

This research was sponsored by DARPA's Strategic Technology Office and conducted within the Acquisition and Technology Policy Center of the RAND National Security Research Division (NSRD), which operates the National Defense Research Institute (NDRI), a federally funded research and development center sponsored by the Office of the Secretary of Defense, the Joint Staff, the Unified Combatant Commands, the Navy, the Marine Corps, the defense agencies, and the defense intelligence enterprise.

For more information on the RAND Acquisition and Technology Policy Center, see www.rand.org/nsrd/atp or contact the director (contact information is provided on the webpage).

Contents

Figures and Table

Figures

Table

Summary

The Defense Advanced Research Projects Agency (DARPA) has an ambitious vision for *Mosaic Warfare*, conceived by DARPA's Strategic Technology Office leadership as both a warfighting concept and a means to greatly accelerate capability development and fielding. Mosaic Warfare takes its name from the concept of assembling individual warfighting platforms—like the small ceramic tiles in mosaics—to make a force package. Although the success of Mosaic Warfare depends on DARPA advancing multiple technologies (Clark, Patt, and Schramm, 2020), the Strategic Technology Office's Mosaic vision is inherently more challenging to transition than is a program or technology. Anticipating this challenge, DARPA sponsored RAND Corporation research to examine the opportunities and challenges associated with developing and fielding a Mosaic force under existing or alternative governance models and management processes, as would be required for the vision to move from DARPA to widespread acceptance by the U.S. Department of Defense (DoD).

A complete survey of the Mosaic warfighting concept is beyond the scope of this report but can be found in other sources (Clark, Patt, and Schramm, 2020; Deptula et al., 2019; Grana, Lamb, and O'Donoughue, 2021; O'Donoughue, McBirney, and Persons, 2021). Briefly, with regard to warfighting, Mosaic Warfare entails a more fractionated, more heterogeneous force that can be dynamically composed on tactical timelines into unique force packages to surprise and overwhelm an adversary. As a result, Mosaic Warfare entails shifting away from a focus on monolithic platforms, which are slow to develop and slow to field, to a focus on simpler force elements that can be developed and fielded quickly and integrated at mission execution. At the top level, the Mosaic Warfare concept envisions a U.S. force characterized by three properties: fractionation, heterogeneity, and composability (O'Donoughue, McBirney, and Persons, 2021). *Fractionation* refers to the extent to which the capabilities of a military force are concentrated on particular weapon platforms. *Heterogeneity* refers to the extent to which the platforms in a military force possess distinct capability sets. *Composability* refers to the extent to which force elements can be dynamically combined in different ways to deliver an operational effect.

For the purposes of the study, we were principally interested in the implications of Mosaic Warfare for requirements, resourcing, and acquisition. In that context, two hypotheses framed our research:

1. DARPA's vision of Mosaic Warfare can enable orders of magnitude of reduction in time for the transition from idea to effect, allowing force development on operational, if not tactical, timescales.
2. Mosaic Warfare might be necessary but will not be sufficient to achieve such increased throughput—it must be complemented with new approaches to setting requirements, resourcing, and acquisition.

We distilled these hypotheses into two research questions:

1. Are DoD's existing requirements, resourcing, and acquisition structures and processes compatible with fielding DARPA's vision of Mosaic Warfare? Are those management systems compatible with the envisioned increases in time-effectiveness?
2. If DoD's current governance systems are not adequate to handle the increased time-effectiveness, what are viable alternative governance models and management systems for acquiring a Mosaic force? What are the opportunities, challenges, and risks associated with them?

To answer these questions, we reviewed existing studies, spoke to experts, and designed and executed the *Acquiring a Mosaic Force Policy Game* to immerse DARPA representatives and RAND researchers in the task of fielding a Mosaic force. We required them to operate with the authorities, responsibilities, and constraints provided to them under existing governance models and management constructs or alternative models and constructs. In total, we executed two internal (RAND-only) play-tests of the policy game while hypotheses and the game design were still in development and one "capstone" game with combined DARPA and RAND participation once these hypotheses were firmer.

Insights About Mosaic Warfare Under the Department of Defense's Current Acquisition Paradigm

Within the existing acquisition paradigm, the logic of Mosaic Warfare can be expected to promote faster, cheaper, more-responsive acquisition at the "tile" level. For example, if we assume that individual Mosaic tiles are represented as programs in a traditional acquisition paradigm, Mosaic Warfare envisions simpler programs through the logic of fractionation. Because program complexity is a known driver of risk, we can expect that, holding other factors constant, greater program-level simplicity will reduce cost, schedule, and performance risk for the acquisition and sustainment of those programs. Similarly, Mosaic Warfare's logic of composability—dynamically combining force elements on tactical timelines—defers a portion of systems engineering and integration from the level of individual tiles to the level of the Mosaic force. Given the known contribution of systems engineering and integration to program risks, we can also expect this feature of Mosaic Warfare to reduce cost, schedule, and performance risk at the tile level.

Neither our exploratory research nor gaming revealed anything inherent within DoD's existing requirements, resourcing, and acquisition systems that would prohibit or even inhibit the development, procurement, operation, or sustainment of forces that are more fractionated, more heterogeneous, and more composable. DoD's management systems and governance structures are by nature technology- and capability-agnostic, and much has been written about "tailoring" the process to match the program and circumstances at hand (e.g., McKernan, Drezner, and Sollinger, 2015).

However, as a general rule, the mechanisms at DoD's disposal to accelerate delivery of capabilities are intended to be exceptions reserved for a small number of requirements or programs (e.g., Urgent Operational Needs). Yet Mosaic Warfare entails rapid acquisition as the rule, not an exception. We found that DoD's existing governance models and management systems likely do not scale: That is, even if DoD could rapidly and cost-effectively field *individual* systems that are more heterogeneous, more fractionated, or more composable, existing governance models and management systems likely would not support fielding and continuously upgrading an entire Mosaic force on operational timescales.

Insights About Mosaic Warfare Under an Alternative Acquisition Paradigm

Because our exploratory analysis indicated a multitude of barriers associated with the existing acquisition system, we embraced the task of formulating a credible alternative as a creative exercise informed by the subject-matter expertise of the team and our DARPA sponsors. To guide our thinking, we set forth four broad principles to shape the formulation of an alternative for DoD to manage risks and allocate resources while attempting to be more responsive to a Mosaic vision:

- Principle 1: Acknowledge DoD's enduring need for risk management and resource allocation.
- Principle 2: Consolidate authority for requirements, resourcing, and acquisition.
- Principle 3: Promote oversight and protect institutional equities by limiting scope.
- Principle 4: Embrace mission centrality.

Using these principles, we codified an alternative that is embodied by a new Joint Mission Office (JMO) that reports to the Secretary of Defense. The JMO consolidates requirements, resourcing, and acquisition oversight into a single office, but, following our third principle, the scope of the JMO's authority is limited to a specific mission (anti-surface warfare), a specific theater (U.S. Indo-Pacific Command [INDOPACOM]), a specific capability (enablers, such as munitions, sensors, and command and control nodes), and specific forces (assigned forces in INDOPACOM). The proposed model also features a governance concept that requires the joint approval of the JMO, the relevant service, and the combatant command (COCOM) before any new capability is fielded, effectively giving veto power to all three parties. We make no upfront claims about the optimality of our concept because this phase was the first in our iterative gaming approach to exploring Mosaic acquisition.

Game players felt that the alternative system provided them with more flexibility than the existing system across the areas of requirements, resourcing, technology transition, acquisition oversight, source selection and contracting, systems engineering and integration, test and evaluation (T&E), and fielding and sustainment. However, the proposed organization was not a cure-all—new risks and challenges were also highlighted, suggesting the value of further analysis of organizational alternatives. For example, the proposed model overcame the barriers in the traditional model that a requirement must precede resourcing and overcame the barrier inherent in the Planning, Programming, Budgeting, and Execution process's two-year budget cycle for clairvoyance on future needs, but the new model introduced a new question of how DoD would garner and sustain support among DoD stakeholders and Congress without a requirement to serve as an agreed-upon benchmark for progress. For another example, the new model was seen as facilitating technology transition by incentivizing a broader view of the S&T pipeline, but questions were raised about whether accelerating the pipeline at scale would exceed the services' and COCOMs' finite capacity to integrate new capabilities into training, sustainment, planning, and other supporting processes. For a final example, the model overcame a barrier of focusing acquisition oversight on programs rather

xvi   Acquiring a Mosaic Force

than joint missions; however, it was noted that this could require acquisition oversight of many programs that may have evaded attention in the existing model that uses cost (e.g., ACAT designation) as its primary indictor of risk. Other examples are discussed in the body of this report. Overall, we do not foresee a single "best" solution; rather, our analysis foreshadows options and trade-offs between enabling Mosaic Warfare, balancing service and COCOM equities under Title 10, and the actual and political cost of the institutional changes necessary to realize a given alternative. For future consideration, we discuss several other alternative models for requirements, resourcing, and acquisition under a Mosaic construct and speculate on trade-offs among them.

Recommendations

We suggest that DARPA continue to experiment with alternative governance systems and management systems. We advise that DoD embrace the following principles when considering such systems:

- **Acknowledge enduring DoD needs for management controls for risk management and resource allocation.** Whatever barriers might exist within today's system; the Joint Capabilities Integration and Development System; Planning, Programming, Budgeting, and Execution; and the defense acquisition system allow DoD to manage risks and responsibly execute hundreds of billions of dollars each year. The fundamental need for risk management and resource allocation will endure even under Mosaic Warfare.
- **Acknowledge service and COCOM equities via Title 10.** Our analysis did not focus on the compatibility between Title 10 and Mosaic Warfare, but we found nothing to suggest any incompatibility. However, any proposed governance model must fully consider service and COCOM equities in legislation.
- **Embrace mission centrality in requirements, resourcing, and acquisition.** An important theme that recurred among players and the research team was the need to elevate oversight and measures of effectiveness above the program to the mission level. We

did not arrive at a final conclusion. However, there appears to be wide consensus among participants about the importance of making mission centrality a focus of Mosaic Warfare. This aligns well with recent Under Secretary of Defense for Acquisition and Sustainment and Under Secretary of Defense for Research and Engineering initiatives to bring a mission engineering–based capability portfolio-management lens to acquisition oversight.

- **Embrace throughput (time-effectiveness) as a Mosaic Warfare measure of merit.** The Mosaic Warfare priority of decreasing the amount of time until fielding envisions Mosaic Warfare as a pipeline of continuously evolving capabilities. Management of such a system requires metrics that embrace the acquisition system as a process without a defined end state. Thus, such a metric as throughput—or time-effectiveness—to reflect the marginal increases in mission effectiveness per unit of time seems an appropriate alternative to program cost, schedule, and performance.

- **Define a measure of merit for Mosaic Warfare that embraces uncertainty.** A core Mosaic Warfare value proposition is adaptability—allowing DoD to adapt on both force development and operational timelines. A measure of merit is needed that captures the robustness of Mosaic Warfare to uncertainty. A body of prior work on robust decisionmaking (Lempert, Popper, and Bankes, 2003) and assumption-based planning (Dewar et al., 1993) could help conceptualize and measure the robustness of Mosaic Warfare to uncertainty.

- **Develop a simulation of the Mosaic pipeline and use it to identify policy levers and bottlenecks that would inhibit realization of a Mosaic force.** Our work employed a policy game to examine issues and trade-offs. It highlighted the important role of managing a capability pipeline with trade-offs between fielding a capability today, investing in the future, incentivizing innovation and competition, not overwhelming operators, and other priorities. Such a pipeline is amenable to modeling and simulation of Mosaic tiles through an acquisition pipeline. The model appears necessary to understand bottlenecks given the potentially large number of small tiles that are wending their way from the laboratory to the field.

Closing Thoughts

DARPA's vision of Mosaic Warfare is ambitious, compelling, and seemingly responsive to many attributes of the emerging technological and security environment. Transitioning this vision to widespread DoD acceptance might require strong proponents across DoD to create change within institutions that today might, in certain cases—given their accrued equity in longstanding governance structures—view the status quo as an end rather than a means. We advise the proponents of Mosaic Warfare to be mindful of falling into the same trap by making Mosaic Warfare an end rather than a means. Like all emerging visions for the future of U.S. warfighting, the ultimate test for Mosaic Warfare will be its contribution to the U.S. ability to deter and defeat adversary aggression.

Acknowledgments

We gratefully acknowledge Tim Grayson and Daniel Javorsek of the Defense Advanced Research Projects Agency's Strategy Technology Office for their intellectual engagement and sponsorship of our study. External game participants provided insights that strengthened our study. Through conversations and trial runs of the acquisition policy game, substantive contributions were made by several RAND Corporation researchers, including Chad Ohlandt, Scott Comes, Irv Blickstein, Maria McCollester, Michael Vasseur, Megan McKernan, Brendan Toland, Brad Martin, William Shelton, and Elaine Simmons.

Abbreviations

ACAT	Acquisition Category
AI	artificial intelligence
ASuW	anti-surface warfare
C2	command and control
CAPE	Cost Assessment and Program Evaluation
CBA	Capabilities-Based Assessment
COCOM	combatant command
DARPA	Defense Advanced Research Projects Agency
DAS	defense acquisition system
DoD	U.S. Department of Defense
DoDD	Department of Defense Directive
DoDI	Department of Defense Instruction
ELINT	electronic intelligence
EOC	essential operational capability
EW	electronic warfare
FAR	Federal Acquisition Regulation
FFRDC	federally funded research and development center

IOC	initial operating capability
INDOPACOM	U.S. Indo-Pacific Command
JCIDS	Joint Capabilities Integration and Development System
JMO	Joint Mission Office
MCC	mission capability compiler
OSD	Office of the Secretary of Defense
OT&E	operational test and evaluation
OTA	Other Transaction Authority
PACOM	Pacific Command
PPBE	Planning, Programming, Budgeting, and Execution
R&D	research and development
RF	radio frequency
S&T	science and technology
sUUV	small unmanned underwater vehicle
T&E	test and evaluation
TTPs	tactics, techniques, and procedures
UAS	unmanned aircraft systems
UAV	unmanned aerial vehicle
USAF	U.S. Air Force
USD(A&S)	Under Secretary of Defense for Acquisition and Sustainment
USD(R&E)	Under Secretary of Defense for Research and Engineering

USMC	U.S. Marine Corps
USN	U.S. Navy
UUV	unmanned underwater vehicle
XLUUV	extra-large unmanned underwater vehicle

CHAPTER ONE

Introduction

The Defense Advanced Research Projects Agency (DARPA) has an ambitious vision for *Mosaic Warfare*, conceived by DARPA's Strategic Technology Office leadership as both a warfighting concept and a means to greatly accelerate capability development and fielding. Mosaic Warfare entails a more fractionated, more heterogeneous force that can be dynamically composed—like the small ceramic tiles in mosaics—on tactical timelines into unique force packages to surprise and overwhelm an adversary. As a result, Mosaic Warfare entails shifting away from a focus on monolithic platforms, which are slow to develop and slow to field, to a focus on simpler force elements that can be developed and fielded quickly and integrated at mission execution. Although the success of Mosaic Warfare depends on DARPA advancing multiple technologies (Clark, Patt, and Schramm, 2020), the Strategic Technology Office's Mosaic vision is inherently more challenging to transition than is a program or technology. Anticipating this challenge, DARPA sponsored RAND Corporation research to examine the opportunities and challenges associated with developing and fielding a Mosaic force under existing or alternative governance models and management processes, as would be required for the vision to move from DARPA to widespread acceptance by the U.S. Department of Defense (DoD).

A complete survey of the Mosaic warfighting concept is beyond the scope of this report but can be found in other sources (Clark, Patt, and Schramm, 2020; Deptula et al., 2019; Grana, Lamb, and O'Donoughue, 2020; O'Donoughue, McBirney, and Persons, 2021). For the purposes of this research, we accept without question the viability and virtues of Mosaic Warfare from a technological and operational perspective so that we may focus on its acquisition-related implications. This report is focused on the implications of Mosaic Warfare for "Big A" acquisition. That is, we are principally concerned with how existing or alternative processes and public management systems for setting requirements, resourcing, and buying weapon systems might effectively equip a Mosaic force. In that context, two DARPA working hypotheses frame our research:

1. DARPA's vision of Mosaic Warfare can enable orders of magnitude of reduction in time for the transition from idea to effect, allowing force development on operational, if not tactical, timescales.

2. Mosaic Warfare might be necessary but will not be sufficient to achieve such throughput—it must be complemented with new approaches to setting requirements, resourcing, and acquisition.

As we will discuss in more detail later, the first hypothesis envisions acquisition as the process that begins conceptually, with an idea on an engineer's whiteboard (perhaps to meet an operational need), and concludes with an operational effect delivered; Mosaic Warfare is envisioned as a means to accelerate an "end-to-end" acquisition process, not merely the delivery of acquisition programs. The second hypothesis presumes that those achievable speeds at scale will require changes to the management and governance systems that DoD employs; in other words, the Joint Capabilities Integration and Development System (JCIDS); Planning, Programming, Budgeting, and Execution (PPBE); and the defense acquisition system (DAS) are inadequate to fully exploit the vision of Mosaic Warfare. Both hypotheses embed assumptions about the viability of the extant systems and the virtues of unstated alternatives. Both working hypotheses are subject to analysis in this report.

Research Questions

We distilled our hypotheses into two research questions:

1. Are DoD's existing requirements, resourcing, and acquisition systems compatible with fielding DARPA's vision of Mosaic Warfare? Are those management systems compatible with the envisioned increases in time-effectiveness?
2. If DoD's current governance systems are not adequate to handle the increased time-effectiveness, what are viable alternative governance models and management systems for acquiring a Mosaic force? What are the opportunities, challenges, and risks associated with them?

Research Approach

Our approach to examining these questions was multifold. We began by conducting exploratory research to immerse the RAND team in DARPA's vision for Mosaic Warfare and in the known costs, benefits, and risks of the existing requirements, resourcing, and acquisition system. This exploration took the form of an examination of relevant literature, supplemented by interviews with retired DoD leaders and staff who are familiar with DoD's existing approach to requirements, resourcing, acquisition, sustainment, and operations. This exploratory research was synthesized into a set of hypotheses about the viability of the extant management systems to support, if not fully exploit, the Mosaic vision. Because of apparent challenges with the existing requirements, resourcing, and acquisition system, the exploratory research yielded an initial set of alternative governance models and management constructs.

To further explore these hypotheses and alternatives, we designed and executed a policy game that immersed the RAND research team and DARPA representatives in the task of fielding a Mosaic force and required them to operate with the authorities, responsibilities, and constraints provided to them under existing or alternative governance models and management constructs. The premise of the exercise was that some of the unintended first- and second-order implications of competing governance models and management principles are more-effectively illuminated experientially than through reliance on the deductive logic of analysts. As is often the case in gaming, the process of designing the exercise also yielded insights. Thus, we adopted an iterative process of formulating hypotheses and then using successive iterations of the policy game to test, revise, and retest these hypotheses. In total, we executed two internal (RAND-only) play-tests of the policy game while hypotheses and the game design were still in development and one "capstone" game with combined DARPA and RAND participation once these hypotheses were firmer.

Finally, all of the phases of the research were integrated to produce a set of insights and next steps for DARPA. This straightforward research approach is summarized in Figure 1.1.

Organization of This Report

The remainder of this report is organized as follows. In Chapter Two, we provide an overarching analytic framework that defines *Mosaic Warfare* and *acquisition* for the purpose of our investigation. Chapter Two also articulates the two acquisition governance models that were the focus of our gaming. Chapter Three then summarizes the design of an acquisition policy game used to explore acquisition in a Mosaic paradigm. Chapters Four and Five summarize the insights gleaned from the policy game and the project as a whole. Chapter Four contains insights related to the relationship between Mosaic Warfare

Figure 1.1
Research Approach

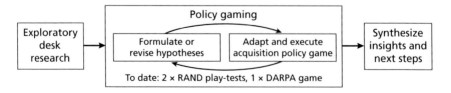

and the existing "Big A" acquisition system, while Chapter Five contains insights related to Mosaic Warfare and the proposed alternative acquisition model. In Chapter Six, we briefly propose three additional acquisition models and propose a trade space relating the cost of implementation and the expected effectiveness of the proposed alternative acquisition apparatus. Chapter Seven summarizes conclusions and next steps.

CHAPTER TWO

Frameworks, Assumptions, and Acquisition Models for Analysis

This chapter synthesizes our exploratory research by presenting working frameworks for conceptualizing Mosaic Warfare and acquisition and describing some broad assumptions that are necessary for our work to proceed. Naturally, DARPA has devoted considerable energy to defining Mosaic Warfare, and the acquisition system is the subject of several decades of policy research. We draw on this prior work and thinking, but our frameworks are not presented to be definitive; rather, they serve the practical purpose of guiding our research and providing more background to contextualize the policy game and insights in the subsequent chapters. We also describe the two acquisition governance models that are the focus of our gaming.

Conceptualizing *Mosaic Warfare*

The Vision

As mentioned, Mosaic Warfare is conceived by Strategic Technology Office leadership as both a warfighting concept and a means to greatly accelerate capability development and fielding. At the top level, the Mosaic concept envisions a U.S. force characterized by three properties:

- *Fractionation* refers to the extent to which the capabilities of a military force are concentrated on particular weapon platforms. A monolithic or nonfractionated force locates a large number of capabilities on one platform. The F-35 is perhaps the canonical example of a monolithic platform; it has the capabilities of a sensor, shooter, command and control (C2) node, electronic warfare (EW), and others all integrated on a single platform. In contrast, a fractionated force spreads such functions and capabilities across an array of platforms. Mosaic Warfare envisions a more fractionated U.S. force.

- *Heterogeneity* refers to the extent to which the platforms in a military force possess distinct capability sets. In a homogeneous force, platforms have a high degree of capability overlap. As DoD transitions away from legacy fourth-generation fighters to the F-35, the U.S. tactical air support fleet will, by definition, grow more homogeneous, notwithstanding differences between F-35 variants and what will be an ever-evolving series of incremental capability upgrades. In a heterogeneous force, platform capabilities will have less commonality and more diversity; for example, the same EW effect might be delivered by an unmanned aerial vehicle (UAV), an aerostat, or a low-cost cruise missile. Mosaic Warfare envisions a more heterogeneous U.S. force.

- *Composability* refers to the extent to which force elements can be dynamically combined in different ways to deliver an operational effect. A highly noncomposable force would be constrained to fixed, prespecified kill chains embodied by a codified system architecture; the ballistic missile defense system represents an archetype. A highly composable force eliminates the concept of an architecture, allowing kill chains to be created dynamically from the force elements available at the time of mission execution. Mosaic Warfare envisions a more composable force; an artificial intelligence (AI)–enabled decision aid will facilitate the force package composition function at the time of mission execution.

For the purposes of this report, we assume that a force with these properties is militarily advantageous and technically feasible. However, we will briefly comment on the operational and acquisition-related advantages of Mosaic Warfare as conceived by DARPA.

In terms of operations, Mosaic Warfare proponents expect that a fractionated, heterogeneous, and composable force will increase the adaptability, scalability, and unpredictability of the U.S. force. The current force, it is argued, comprises force packages that are self-contained or part of fixed system-of-systems architectures and thus limited in terms of the distinct force presentation permutations available. In contrast, a Mosaic force will decompose force packages into a larger number of more-varied elements, thereby increasing the number, the resiliency, and, ultimately, the effectiveness of force packages available for employment by U.S. commanders. Clark, Patt, and Schramm, 2020, p. 27, succinctly characterizes some of the hypothesized warfighting advantages of Mosaic Warfare:

> The central idea of the Mosaic Warfare concept is to create adaptability and flexibility for U.S. forces and complexity or uncertainty for an enemy through the rapid composition and recomposition of more disaggregated U.S. forces using human command and machine control.

DARPA also anticipates that Mosaic Warfare might accelerate the weapon system acquisition and fielding process. The development of complex multimission platforms is slow and expensive. Much of the cost and schedule expended in the development of these platforms stems from a requirements system that attempts to forecast general-purpose requirements, which tend to prescribe costly, complex solutions embodied by monolithic solutions. By fractionating systems—and thereby decreasing the average complexity of systems in the acquisition pipeline—Mosaic Warfare is anticipated by DARPA to entail individually simpler systems that are subject to less cost, schedule, and performance risk; defer integration challenges to the mission level; and result in a flexible, modular force that can be continually upgraded over time. Deptula et al., 2019, p. 34, explains that the functional effect of transitioning to a Mosaic force composition on acquisition and fielding might be to realize the benefits sought during the many recent rounds of acquisition reform:

> Incrementally migrating the current force to a system of disaggregated capabilities is an approach that could finally achieve

the goals that many of DOD's previous attempts at acquisition reform have sought.

The elements of a Mosaic force also can be expected to be more autonomous, expendable, and short-lived than the technologies comprising today's force. These traits might positively reinforce the core Mosaic concepts of fractionation, heterogeneity, and composability. For example, autonomous systems might hasten the anticipated speedup in fielding by eliminating certain portions of the operator training cycle. Expendable systems can be expected to eliminate time-intensive sustainment processes, such as repair, maintenance, and upgrading. Shorter weapon system lifespans might obviate the cost and schedule implications of a requirement to maintain long (e.g., 30-year) service lifetimes.

To be sure, this is an abbreviated, if not incomplete, description of DARPA's vision for Mosaic Warfare. However, it suits our purpose of introducing Mosaic Warfare in sufficient detail to motivate several assumptions. We refer the reader to references cited in the preceding discussion for further information on Mosaic Warfare.

Conceptualizing *Acquisition*

We conceive of *acquisition* in a general sense that includes the end-to-end timeline that begins with an idea on an engineer's whiteboard and culminates with an operational effect delivered on the battlefield. Today, DoD exercises management control over this process through three primary management systems: the requirements system, manifested by JCIDS; the resourcing system, represented by PPBE; and the DAS, represented by Department of Defense Directive (DoDD) 5000.01 and Department of Defense Instruction (DoDI) 5000.02 and, more recently, by the Adaptive Acquisition Framework.[1] In this report, we

---

[1]   There are other relevant management systems in play, including the Global Force Management Process, which governs DoD posture and force allocation, and, of course, multiple operational planning processes. In this report, we focus on the systems that are traditionally considered part of the "Big A" acquisition process.

are principally concerned with the composition of the functions contained in these three management systems—sometimes referred to as the "Big A" acquisition system.[2]

An initial task in this project was to develop a functional decomposition of the steps that need to be accomplished in a generic end-to-end acquisition process. Such a functional decomposition is fundamental in the sense that it is independent of its implementation in particular management systems like JCIDS, PPBE, and the DAS. For example, in the current system, the requirements process, instantiated in JCIDS, typically requires that an operational need initiate the acquisition process. That is, the identification of a demand must proceed the initiation of technological development. In the functional decomposition, presented in Figure 2.1, this sequence need not be maintained; a technological breakthrough might fill the identification of an operational use. Unavoidably, the terms that we use to describe these functions come with baggage associated with their current implementation.

Figure 2.1 depicts a set of functions inherent to the "Big A" acquisition system. The figure is not meant to depict the current implementation of DoD requirements, resourcing, and acquisition processes; rather, it is a depiction of the *functions* performed by the current apparatus, situated left to right in an approximation of when in the capability development life cycle each function takes place. A function's vertical location in the figure is not meant to convey information about the function.

Several features of the functions depicted in the figure are worth highlighting. First, our functional decomposition suggests that acquiring a weapon system requires the completion of a large number of com-

---

[2]   We do not provide a comprehensive summary of the status quo acquisition system in this report. The current (as of March 2021) DoD guidance on these systems consists of the following: *Implementation of the Joint Capabilities Integration and Development System* (Chairman of the Joint Chiefs of Staff Instruction 5123.01H, 2018); *Department of Defense Financial Management Regulation* (Department of Defense 7000.14-R, 2019); *The Defense Acquisition System* (DoDD 5000.01, 2020); *The Planning, Programming, Budgeting, and Execution (PPBE) Process* (DoDD 7045.14, 2017); *Operation of the Adaptive Acquisition Framework* (DoDI 5000.02, 2020); *Operation of the Defense Acquisition System* (DoDI 5000.02T, 2020); and *Manual for the Operation of the Joint Capabilities Integration and Development System* (DoD, 2018).

Figure 2.1
A Space of Functions in Defense Acquisition

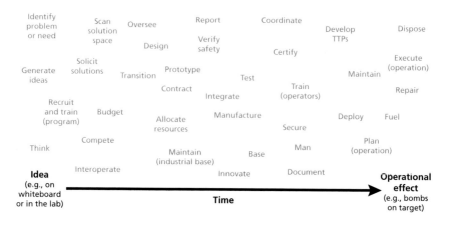

NOTE: TTPs = tactics, techniques, and procedures.

plex functions. Although the process of attaining effective and needed weapon systems likely can be simplified, it is unlikely ever to be simple.

Second, the functions depicted in the figure require very distinct kinds of knowledge and organizational capacity. The requisite expertise to identify an operational problem might have little overlap with that needed to manufacture the technology in question. In practice, in any alternative "Big A" acquisition system, this is likely to necessitate participation by a large number of distinct, geographically dispersed organizations. Thus, end-to-end process efficiency will depend not only on the acuity with which each function is undertaken but also on the interorganizational coordination of the distinct functions.

Third, our functional decomposition contains the three major functional categories of the current "Big A" acquisition system: requirements, resourcing, and acquisition. These functions are fundamental to any government-run process of acquiring technologically complex goods and can be defined as follows:

- *Requirements* ("Identify problem or need" in Figure 2.1). Any system designed to equip an organization with the material means of achieving that organization's objectives must have some way of identifying and characterizing the organization's needs and, in a bureaucracy as large as DoD, assigning responsibility for meeting those needs. Functionally, such a needs process affords the means by which the "Big A" acquisition process determines *what* to acquire. Important dimensions on which a needs system can vary are how and when a requirement is articulated. Articulating a need with high specificity early in the acquisition process might streamline the solution search process but prevent a broad exploration of the full solution space. In contrast, articulating a need with low specificity might delay the solution-winnowing process but expand the portion of the solution space that is considered.

- *Resourcing* ("Allocate resources" in Figure 2.1). Any acquisition system that faces a condition of resource scarcity must have some means of determining what to fund. The resourcing function determines what is funded; it specifies the process for allocating funds and how those funds can be used. One design decision faced by any resourcing system relates to the fungibility of allocated funds. High fungibility—the ability to rapidly switch the intended use of allocated funds—allows organizations to adjust resource allocations when conditions change. The cost of this flexibility is reduced pre-allocation control over how funds are used. Any system characterized by low resource fungibility will limit an organization's ability to adjust to changed circumstances but afford more predictability and management control.

- *Acquisition* (many functions in Figure 2.1). In addition to a means of identifying demand and determining how and when to fund an acquisition, any acquisition system requires a set of processes and an associated organizational apparatus to administer the various stages of a government-run development effort and purchase decision. Functions that fall into this category include scanning the solution space, soliciting solutions, managing the vendor bids process ("Compete" in the figure), contracting, and testing.

Finally, it is worth highlighting that our functional decomposition contains tasks that sit outside the current functional domain of the acquisition community as it is traditionally conceived of. For example, in the current system, the tasks of developing TTPs and manning weapon systems largely fall to the combatant commands (COCOMs). Considering the full complement of processes that affect acquisition might require expanding the scope of organizations and functions that are considered to be *within* the acquisition community.

Assumptions

As mentioned, for the purposes of this research, we accept without question the viability and virtues of Mosaic Warfare from a technological and operational perspective so that we may focus on its acquisition-related implications. More specifically, we specify several assumptions:

1. **Technical interoperability between and within Mosaic elements is seamless.** A highly heterogeneous, fractionated, and composable force requires a high degree of interoperability across Mosaic elements. This is a significant challenge, but it is largely a technical one. To focus on the implications of a Mosaic force for DoD resourcing, requirements, and acquisition processes, we assume that the technical interoperability issue has been resolved. Multiple DARPA programs are examining means to achieve necessary levels of technical interoperability.

2. **A mission capability compiler (MCC) has been advanced and demonstrated to a degree that its recommendations are viewed as credible by DoD leadership, staff, and their U.S. government stakeholders.** A Mosaic force relies on AI to compose individual elements into kill chains at the time of mission execution and to guide force development and design with recommendations on the quantity and quality of new capabilities to develop, acquire, and field. Here, we posit an MCC that provides both of these functions. The MCC is armed with a representation of the existing U.S. capability set and paired with a sufficiently realistic, multiresolution model of the operational environment to evaluate the efficacy of the force in specific missions and conditions. The MCC can thus simulate the perfor-

mance of hundreds, if not thousands, of alternative force compositions in myriad scenarios and develop recommendations for operators or planners on alternative courses of action. The MCC can also recommend to force planners the acquisition of new or improved capabilities that would improve the performance of the Mosaic force. The viability of such a trusted AI-assisted decision aid is at present well beyond the technical state of the art. In this report, we assume that the MCC exists and that its recommendations are credible to relevant DoD stakeholders. At various points during game play, we appealed to advice or estimates provided by the MCC.

3. **OSD, the Joint Staff, the services, and COCOMs will maintain authorities, interests, missions, and top-level priorities, as derived from Title 10.** The game explored alternative allocations of authority and resources for requirements, resourcing, and acquisition. However, we endeavored to maintain a basic commitment to Title 10 such that the posited allocations can be understood as alternative implementations of existing law.

4. **DoD has embraced the Mosaic concept and successfully transitioned to an initial Mosaic force.** The Mosaic vision is not without controversy, and the premise of our study is that DoD will need to adapt to adopt Mosaic Warfare at scale. We assume widespread support for the Mosaic warfighting concept. DoD stakeholders might have different priorities and maintain potentially divergent institutional interests, but we assume that their priorities and interests—and resulting bureaucratic positions—are shaped within an agreed-upon framework of a fractionated, heterogeneous, and composable force as embodied by Mosaic Warfare. Put starkly, the institution might disagree on *which* fractionated, heterogeneous, and composable force to procure—but not that a fractionated, heterogeneous, and composable force *should* be procured.

5. **The scope of Mosaic acquisition is limited to enablers (such as sensors, munitions, C2 nodes, and attritable platforms[3]), leaving the task of acquiring major platforms to the services.** Mosaic Warfare does not require the wholesale replacement of all major weapon platforms, monolithic or otherwise. We assume that a need remains for monolithic fighter aircraft (e.g., F-35), heavy bombers (e.g., B-21), attack submarines (e.g., *Virginia*-class submarines), and other large platforms and that the existing institutional processes and policies for developing, funding, acquiring, and sustaining these platforms will persist. We assume that the scope of Mosaic Warfare is limited to enablers, such as sensors, communication nodes, munitions, and expendable/attritable platforms, which might not be hosted on major platforms that the services field. In all governance alternatives considered in this report, the services will maintain responsibilities for acquiring major platforms, with requisite oversight by the Office of the Secretary of Defense (OSD) and the Joint Staff.

Models for Analysis

The Baseline Model

Our policy game considers the extant DoD requirements, resourcing, and acquisition paradigm as the baseline for analysis. These systems are the subject of entire courses within the Defense Acquisition University and are the focus of decades of policy research. Therefore, we eschew a detailed description of them here, except to say that they are represented broadly by JCIDS, PPBE, and the DAS, including the new pathways offered by the Adaptive Acquisition Framework, such as the Middle Tier of Acquisition pathway.

---

[3] A weapon system can be attritable if, by virtue of low unit cost, the system's loss is considered relatively tolerable (Colombi et al., 2017).

An Alternative Model

Our exploratory analysis indicated a multitude of barriers associated with the existing system, requiring us to confront our second research question: What are viable alternatives? We embraced the task of formulating a credible alternative as a creative exercise informed by the subject-matter expertise of the team and our DARPA sponsors. To guide our thinking, we set forth four broad principles to shape the formulation of an alternative for DoD to manage risks and allocate resources while attempting to be more responsive to a Mosaic vision. We make no upfront claims about the optimality of our vision or principles because this phase was the first in our iterative gaming approach to exploring Mosaic acquisition.

Principle 1. Acknowledge the Department of Defense's Enduring Need for Risk Management and Resource Allocation

In the process of exploring alternatives, what quickly came into focus was a need to understand what is truly fundamental about the existing paradigm. How can barriers in the current system be addressed without throwing the proverbial baby out with the bathwater? Whatever barriers might exist within today's system, JCIDS, PPBE, and DAS perform essential functions. JCIDS provides a way to identify, prioritize, and assign responsibility for DoD needs. PPBE provides means to allocate, adjudicate, and synchronize investments across services, programs, and time frames. The DAS oversees cost-schedule-performance outcomes in execution and the overall health of the industrial base. The fundamental need for demand identification, resource allocation, and risk management will endure even under Mosaic Warfare. Furthermore, these functions are inherently political: Even under Mosaic Warfare, DoD will require means for institutional stakeholders to express and resolve often-competing interests. Our first principle is to acknowledge the essential functions performed by the existing acquisition apparatus and to ensure that these functions are retained in any proposed alternative.

Principle 2. Consolidate Authority for Requirements, Resourcing, and Acquisition

A selected review of attempts to greatly accelerate DoD acquisition reveals a common feature: the consolidation of requirements, resourcing, and acquisition authority. Such organizations as the Missile Defense Agency, the Joint Improvised Explosive Device Defeat Organization, the Strategic Capabilities Office, Special Operations Command, and service rapid capabilities offices, among others, were empowered to shape their own requirements, were able to flexibly allocate resources to meet them, and were given authority to execute. To be clear, we do not claim these examples to be unmitigated successes, and there are important differences between their respective mandates, implementation, and outcomes. Nonetheless, this seemingly common feature motivated us to consider an alternative that concentrated authority for requirements, resourcing, and acquisition.

Principle 3. Promote Oversight and Protect Institutional Equities by Limiting Scope

The consolidation of power creates a corresponding need for effective oversight. It also raises concerns about the potential challenge of resolving the interests of competing (if not newly weakened) institutions. We were reminded of the proverb, "Absolute power corrupts absolutely." Our third principle for designing an alternative system is to greatly limit the scope of responsibility as a way of bounding the unintended consequences of concentrated power.

Principle 4. Embrace Mission Centrality

Our fourth principle is to embrace mission centrality. Today's requirements, resourcing, and acquisition systems conceptualize programs as the central organizing element. We do not dispute the validity of a program as a construct: Ultimately, someone somewhere will be responsible for managing the activity of a specific capability. However, we sought in our approach to elevate the prominence of the mission in acquisition, such that oversight of requirements, resourcing, and acquisition is managed at the mission level rather than the program level. As an aside, this appears consistent with the ongoing efforts of the Under Secretary of Defense for Acquisition and Sustainment (USD[A&S])

and the Under Secretary of Defense for Research and Engineering (USD[R&E]) to employ mission engineering–based capability portfolio management to structure acquisition oversight (Office of the Deputy Director for Engineering and Office of the Under Secretary of Defense for Research and Engineering, 2020).

The Concept: A Joint Mission Office

Using these principles, we codified an alternative that is embodied by a new Joint Mission Office (JMO) that reports to the Secretary of Defense and is led by a three- or four-star officer with a history of joint service. The JMO consolidates requirements, resourcing, and acquisition oversight into a single office, but, following our third principle, the scope of the JMO's authority is limited to a specific mission (antisurface warfare [ASuW]), a specific theater (U.S. Indo-Pacific Command [INDOPACOM]), a specific capability (enablers, such as munitions, sensors, C2, communication nodes), and specific forces (assigned forces in INDOPACOM). The proposed model also features a governance concept that requires the joint approval of the JMO, the relevant service, and the COCOM before any new capability is fielded, effectively giving veto power to all three parties.

In our concept, the JMO is responsible for publishing an annual prioritized list of ASuW capability needs; assessing all service investments in ASuW capabilities in the Western Pacific; and advising the Secretary of Defense on programmatic changes to support Mosaic Warfare. The JMO also is responsible for managing and competitively allocating its supplementary funds (from its budget) for services or agencies to develop, produce, or sustain ASuW enablers. Furthermore, the JMO is responsible for resourcing and managing a joint ASuW test and evaluation (T&E) range and virtual ASuW T&E environment and for establishing, promulgating, and ensuring service compliance with Mosaic compatibility policy. We allow the JMO to, by exception, initiate and oversee new research and development (R&D) programs for ASuW Mosaic enablers.

In this proposed alternative acquisition model, the JMO is responsible for establishing, managing, and administering contracts within a preapproved DoD-wide vendor pool. Although such a vendor pool may take many forms, we propose the following as one option. In the Mosaic vendor pool, participants are granted four-year contracts, which are awarded via a competitive process. Market participation contracts are renewable at the discretion of the government, and all of the participants have security clearance for the duration of the contract (i.e., not on a per-project basis). The market entry process opens every two years to allow a regular inflow of innovative organizations. There are few restrictions on market participation; the market comprises traditional defense-servicing firms, nontraditional vendors, military service laboratories, university research centers, and federally funded research and development centers (FFRDCs). In addition to allowing participation by organizations specializing in direct capability development, the market allows participation by innovation intermediaries, such as venture capital firms and individuals capable of coordinating an effort to develop a solicited capability. Market participants receive an annual fixed payment to conduct relevant R&D at the direction of DoD.[4] This R&D tasking occupies market participants during periods of latency and is focused on the development of Mosaic architectures, applied research, and prototype development. In addition to the annual fixed payment, participants may be awarded additional contracts for particular development initiatives.

In this model, services retain full responsibility (under Title 10) for development, production, operation, and sustainment of non-enablers (e.g., major platforms) and, by default, retain responsibility to *execute* development, production, operation, and sustainment of enablers. They also retain responsibility for defining "-ility" requirements for all programs.

In this model, the COCOM, in conjunction with the ASuW JMO, develops concepts of operation, concepts of employment, and TTPs for capabilities acquired by the JMO. Jointly with the JMO and the services, the appropriate COCOM approves new ASuW capabilities for fielding to assigned forces.

---

[4]   Participants that are intermediaries—that lack an in-house R&D capability—do not receive this portion of participant remuneration.

To facilitate our policy gaming, we created a mock DoDI that codifies this proposed governance structure using language representative of how it might appear in an official communique. Figure 2.2 depicts the faux DoDI.

Figure 2.2
A Faux Department of Defense Instruction

 Faux Department of Defense Instruction – Page 1
ASuW JMO

Mission: The ASuW JMO shall ensure the Joint Force's continual ability to execute the ASuW mission set given a dynamic threat environment and evolving space.

The ASuW JMO shall, inter alia:
- Publish an annual prioritized list of ASuW capability needs.
- Assess all Service investments in ASuW capabilities in the Western Pacific with regard to Mosaic Readiness.
- Advise the Secretary of Defense annually on programmatic changes to support Mosaic Warfare.
- Manage and allocate, via a competitive process, a fund to support the Services or Defense Agencies to develop, produce, operate, or sustain ASuW enablers.
- Resource and manage a Joint ASuW test and evaluation (T&E) range and virtual ASuW T&E environment.
- Establish, promulgate, and ensure Service compliance with Mosaic compatibility policy.
- Establish, manage, and administrate contracts of a preapproved vendor pool.
- By exception, initiate and oversee new R&D programs for ASuW Mosaic enablers.
- Initiate, set terms, and select winners of competitions in Mosaic prize competitions.

 Faux Department of Defense Instruction – Page 1
The Institution: Services, COCOMs, etc.

The Secretary of Defense shall:
- Determine, in conjunction with the Deputy Secretary of Defense and the Vice Chairman of the Joint Chiefs of Staff, the high-level mission set of the ASuW JMO.
- Appoint Directors of JMO based on CJCS nominations.

The Services shall, inter alia:
- Retain full responsibility (under Title 10) for development, production, operation, and sustainment of nonenablers.
- By default, retain responsibility to execute development, production, operation, and sustainment of enablers.
- Retain responsibility for defining "ility" requirements for all programs.
- Jointly with JMO and INDOPACOM, annually approve new ASuW capabilities for fielding to assigned forces.

INDOPACOM shall:
- Create, in conjunction with the ASuW JMO, concepts of operation, concepts of employment, and TTPs for capabilities acquired by JMO.
- Jointly with JMO and Services, approve new ASuW capabilities for fielding to assigned forces.

CHAPTER THREE

The Mosaic Acquisition Policy Game

In this chapter, we briefly describe the acquisition policy game designed to explore the consequences of acquiring a Mosaic force within the current governance model and an alternative model.

Game Design

To better understand how the current governance model and an alternative model would work in conjunction with Mosaic Warfare, we developed a three-part virtual event, depicted in Figure 3.1. Each of the three activities took the form of a virtual half-day session. The activities were conducted two times internally and once with a mixed group of RAND and DARPA personnel. As detailed in Chapter Two, the activities posited that Mosaic Warfare was technically feasible, that

Figure 3.1
Structure of Virtual Event

	Half Day 1: Mosaic Warfare in Today's System	Half Days 2 and 3: Mosaic Warfare in an Alternative Model
Goal of exercise	Identify conditions under which today's requirements, resourcing, and acquisition systems support a Mosaic model	Exercise an alternative to today's management systems to assess viability and identify improvements
Role of participants	Experienced professionals and analysts	Role-playing DoD stakeholders

it was accepted by DoD, and that an initial suite of capabilities had been successfully fielded. From this starting point, we asked players to consider how the acquisition of both individual capabilities and the Mosaic enterprise as a whole might be managed.

Day 1 focused on how the current acquisition system could accommodate both individual tiles (i.e., specific Mosaic enabler technologies) and a Mosaic force as a whole. This activity was designed to explore the shortfalls of the current system by working to identify "pain points." To do this, we used a format based on a previous RAND game designed for assessing C2 structures (Alkire, Lingel, and Hanser, 2018). Drawing on the principles of assumption-based planning (Dewar et al., 1993), we provided players with a set of vignettes (included in the appendix) describing instances of successful Mosaic acquisition, including descriptions of requirements, resourcing, contractor selection and management, T&E, fielding, and maintenance and sustainment. Players were then asked to describe what assumptions would have to hold true in today's system for the vignette to play out as described. A facilitator then led a discussion regarding the reasonableness of those assumptions. This process allowed players to grapple with the difficulty of making Mosaic Warfare work under the current rules and processes, adding to our understanding of the barriers to acquiring a Mosaic force.

The second two activities changed the focus from examining Mosaic acquisition under the current system to exploring it under an alternative system. Although these activities drew on lessons from past work on acquisition policy gaming (Bartels, Drezner, and Predd, 2020), the activities used on days 2 and 3 were designed specifically for exploring acquisition under a Mosaic Warfare construct. This allowed us to explore the interaction between a pipeline of Mosaic capabilities and an alternative acquisition system designed to accommodate Mosaic acquisition.

Entering days 2 and 3, we presented the JMO-centered acquisition model (the details of the JMO as played are described in Chapter Two). The second activity focused on how enterprise-level acquisition management might occur under the JMO-centered model. The third activity focused on tile-level decisions under the same model.

In both activities, players were divided into two teams. One team comprised players representing the ASuW JMO. The other team comprised players representing traditional institutional players: the services, COCOMs, and OSD. A more detailed breakdown of roles is presented in Figure 3.2, which shows players seated around a table. Players were assigned to roles that mirrored their expertise. These roles are shown in Figure 3.3. Using experienced players allowed us to depend on participants' mental models of institutional equities, authorities, and processes to bring additional realism and raise concerns that we might not have been aware of. Thus, our players added greater fidelity to the representation of the interactions between the JMO and the institutional roles.

In the second activity, players were provided with a portfolio of artificial Mosaic programs (i.e., Mosaic tiles) and a budget and asked to make decisions about which tiles to fully fund, which to keep warm for potential future investment, and which to terminate. For each tile,

Figure 3.2
Players Inhabit the Roles of Department of Defense Decisionmakers

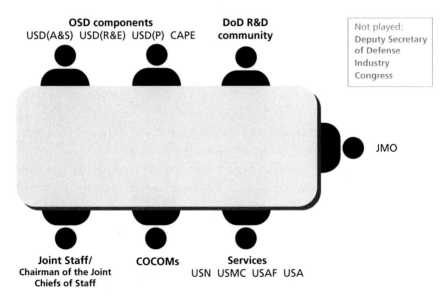

NOTE: CAPE = Cost Assessment and Program Evaluation; USA = U.S. Army; USAF = U.S. Air Force; USD(P) = Under Secretary of Defense for Policy; USMC = U.S. Marine Corps; USN = U.S. Navy.

Figure 3.3
Players' Backgrounds Reflect Assigned Roles

Players in RAND Play-Tests I and II	Players in DARPA Game
Former DoD officials on RAND staff, for example,	• DARPA Strategic Technology Office leadership and staff
• Retired O6, USN representative for JCIDS	• Retired OPNAV N81
• Retired acting director of CAPE	• Former USD(ATL) staff member
• Former USD(ATL) staff member	• Senior adviser to USD(A&S)
• Former USN Director for Analysis, NAVAIR	

NOTE: NAVAIR = Naval Air Systems Command; OPNAV = Office of the Chief of Naval Operations; USD(ATL) = Under Secretary of Defense for Acquisition, Technology, and Logistics.

players were provided with a description of the system and data, including estimates of cost, schedule, and anticipated gain in mission effectiveness. JMO players were asked to use these data to develop a strategy for acquisition, while institutional players acted as liaison officers to represent the concerns of their offices. After a first round of decisions, we projected how the portfolio would perform the following year, with a specific eye toward highlighting the trade-offs identified in previous stages of research.

The third activity maintained the same general structure as the second: The JMO and the institutional teams made sequences of decisions about the acquisition of Mosaic capabilities. However, during the third activity, decisions were made at the level of individual Mosaic tiles. In this activity, we presented players with several key decision points, which were designed to elicit potential points of tension between the JMO and the institutional actors. The resultant discussion provided confirming evidence to support our preliminary hypotheses, unearthed additional tensions, and offered tension mitigation strategies.

Vignettes

The intersection of the Mosaic Warfare vision and acquisition is sufficiently uncertain and abstract that planners and policymakers might benefit from concrete representations of how Mosaic acquisition would manifest. One way of providing this tangibility is through vignettes, which illustrate possible manifestations in narrative form. The vignettes describe successful instances of Mosaic acquisition at the tile and enterprise levels. The tile-level vignettes comprise a set of events associated with the acquisition of a new electronic intelligence (ELINT) sensor, an EW payload, and extra-large unmanned underwater vehicle (XLUUV)–launched loitering unmanned underwater vehicle (UUV) munitions. The enterprise-level vignettes consider Mosaic acquisition at the level of the DoD enterprise. They describe the changes to the force structure, industrial base, and R&D pipeline associated with shifting a portion of acquisition from monolithic platforms to dozens of new short-lived, low-cost Mosaic tiles (e.g., attritable platforms, data links, C2 nodes, decision aids, sensors, loitering munitions, small satellites, and counter–unmanned aircraft systems [UAS] systems). The vignettes do not reference any specific acquisition governance model or management system. Instead, they provide generic descriptions of requirements, resourcing, vendor selection, T&E, fielding, maintenance, and sustainment events for the capabilities in question. In the appendix, we provide a series of such vignettes that were used in our acquisition policy game and that might have broader utility in helping DoD envision and plan for Mosaic Warfare.

CHAPTER FOUR

Insights About Mosaic Warfare Under the Department of Defense's Current Acquisition Paradigm

In this chapter, we discuss a series of insights about acquiring a Mosaic force under the current system that emerged from our exploratory research; the design, prototyping, and exercise of the acquisition policy game; and the subject-matter expertise of our team. We selectively draw anonymous quotes from our game as a way to illustrate certain points using language expressed by the players themselves. Unless otherwise stated, we make no general claim as to whether the quotes represent an agreed-upon position of the group.

Logic of Mosaic Warfare Might Promote Faster, Cheaper, More-Responsive Acquisition at the Tile Level, Regardless of the Acquisition Model

When we assume that individual Mosaic tiles are represented as programs in a traditional acquisition paradigm, Mosaic Warfare envisions simpler programs through the logic of fractionation. Because program complexity is a known driver of risk, we can expect that, holding other factors constant, greater program-level simplicity will reduce cost, schedule, and performance risk for the acquisition and sustainment of those programs.

Furthermore, Mosaic Warfare's logic of composability—dynamically combining force elements on tactical timelines—defers, at least in theory, a portion of systems engineering and integration from the level of individual tiles to the level of the Mosaic force package. Given the known contribution of systems engineering and integration to program risks, we can also expect this feature of Mosaic Warfare to, holding other factors constant, reduce cost, schedule, and performance risk at the tile level. Impacts at the Mosaic level are likely to depend on the specifics of interoperability and thus were not examined in our study.

By enabling faster, cheaper, or more-responsive acquisition, Mosaic Warfare might enable several virtuous cycles that might further improve acquisition performance of these dimensions. For example, faster acquisition means that fielded capabilities are more responsive to the threat. During the process of planning for the near term, correspondence between anticipated capability needs and actual capability needs is likely to be high. However, as the planning period increases, so does uncertainty regarding the future threat environment. Increased uncertainty regarding the threat environment leads rational planners to hedge regarding anticipated capability needs, effectively adding requirements designed for future operational environments and future adversary capabilities.[1] A short temporal gap between threat onset and the availability of a threat-mitigating capability theoretically lessens the risk of "requirements creep" that sometimes manifests when the objectives of delayed programs are expanded to accommodate an evolving threat.

Faster adaptation might also come with an expectation of shorter average weapon system service lives, which permits a relaxation of the need to design and build in complex life-cycle sustainment requirements, again lowering costs and schedule risks. Lower costs and risks might motivate less attention from OSD and Congress, potentially sparing the acquisition system the time and expense of heavy oversight. Simpler requirements might also expand the pool of performers who are qualified to develop or produce the tiles, creating opportunities for

---

[1]   Tim Grayson, director of DARPA's Strategic Technology Office, described this relationship between time and requirements growth in a 2019 interview on the program *Government Matters* (Wagner, 2019).

cost-reducing, performance-enhancing innovation and competition. Of course, these judgments rest critically on the assumptions of technical interoperability, which have been more fully described elsewhere (see Fortunato, 2016; Clark, Patt, and Schramm, 2020).

Although these positive effects of a Mosaic paradigm follow from the logic explained in the preceding paragraphs, they were also anticipated by players immersed in a Mosaic world. For example, reflecting on the cost of an exemplar portfolio of Mosaic tiles, one player commented, "These programs are budget dust; let's just do it." Another player stated,

> If . . . we change the life cycle to be much more short-term, that changes the approach for how long you need to test, how much risk you are willing to take, and the hurdles to implementation that you would face. . . . You might be able to get around the uncertainty issues in a decision to proceed with the program.

Finally, another player remarked on the effect of shorter program life spans in truncating a program's sustainment tail and thus reducing sustainment cost, noting, "The shorter the timeline, the better for sustainment because the less money."

Figures 4.1 and 4.2 synthesize a logic for how the principles of Mosaic Warfare might promote reduced cost, schedule, and performance risk. It should be noted that the logic depicted assumes conditions in which everything else is held constant.

Figure 4.1
A Logic Connecting Mosaic Warfare Principles to Acquisition Outcomes

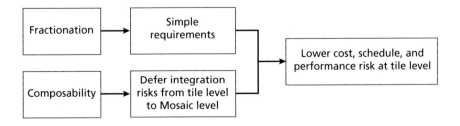

Figure 4.2
Cost- and Risk-Reducing Virtuous Cycles Potentially Enabled by Mosaic Principles

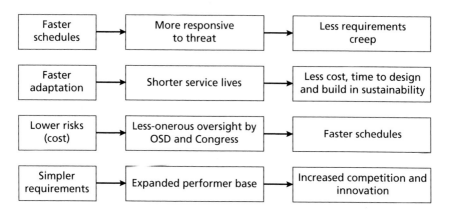

Department of Defense Can Field Individual Capabilities on Mosaic Timescales Today

Neither our exploratory research nor the gaming revealed anything inherent within DoD's existing requirements, resourcing, and acquisition systems that would prohibit or even inhibit the development, procurement, operation, or sustainment of forces that are more fractionated, more heterogeneous, and more composable. DoD's management systems and governance structures are by nature technology- and capability-agnostic, and much has been written about tailoring the process to match the program and circumstances at hand (e.g., McKernan, Drezner, and Sollinger, 2015).

The question that soon emerged, rather, was whether DoD could field such capabilities on the condensed timelines envisioned by Mosaic Warfare. When presented with tile-level vignettes (see vignettes A, B, and C in the appendix), players in our play-tests and in the capstone game cited conditions under which DoD is able to accelerate new capabilities to the field. For example, when prompted about whether and how requirements for the underlying capabilities in the vignettes might be determined, players identified several options, including using Urgent Operational Needs, enlisting the Secretary of Defense or another senior leader to personally champion the effort, or piggy-backing onto an existing priority requirement. The Mine-Resistant Ambush Protected program was cited as an example in which several of those approaches were used.

When prompted with the question of whether resources could be located to fund the accelerated capability development described in the tile-level vignettes, players noted that, in today's paradigm, DoD has several options. These include using fenced pools of funding for rapid fielding (e.g., flexible funding accounts), reallocating resources to the emergent priority in the year of execution by submitting reprogramming requests to Congress, or framing the new tile-level development effort as a sustainment activity within an existing acquisition program.

Finally, when players were prompted to explore the feasibility of accelerating the many acquisition functions in the context of the vignettes, they acknowledged DoD's long-established rapid-fielding and -prototype programs, the more recent but as yet unproven Middle Tier of Acquisition pathway, and the practice of deferring operational T&E (OT&E) by accepting the risk of testing in the field or using Other Transaction Authority (OTA)–type contracting mechanisms, among others.

Recent RAND research highlights these and other means of accelerating capability development and acquisition through the development of an inventory of the many existing means to accelerate acquisition (Anton et al., 2020). This further establishes that DoD should, in principle, be able to accelerate the development and acquisition of fractionated, heterogeneous, and composable systems on Mosaic timescales today.

Current Means of Gaining Speed in Acquisition Are Limited in Scale Under the Current System

As a general rule, the tools discussed in the preceding section are intended to be exceptions reserved for a small number of requirements; however, Mosaic Warfare entails rapid acquisition as the rule, not an exception. Players and team members quickly acknowledged that the bureaucratic solutions that they identified were not standard practice. For example, DoD does not, as a rule, employ Urgent Operational Needs; the Secretary of Defense's bandwidth is finite, precluding the Secretary of Defense from championing all programs; legal requirements commit DoD to testing; and Federal Acquisition Regulation (FAR)–based contracts embed important protections for some programs, among other limitations. Thus, we explored the barriers that might inhibit scaling up the existing model to a Mosaic force.

We found that the existing systems likely do not align with the Mosaic vision of fielding many capabilities on operational timescales. Even if DoD could cost-effectively field systems that are more heterogeneous, more fractionated, or more composable, the amount of time required to execute the functions inherent to fielding is likely incompatible with fielding on those timescales. Although there are ways of accelerating all of the functions inherent to fielding—from contracting to resourcing to T&E—the options appear to be exceptions rather than the rule. Let us consider some of the specific barriers that emerged through our gaming and analysis by considering, in turn, requirements, resourcing, technology transition, acquisition oversight, source selection and contracting, systems engineering and integration, T&E, and fielding and sustainment.

Requirements

Today, JCIDS is the way in which DoD identifies, prioritizes, and assigns responsibility for DoD needs. Our exploratory analysis and policy gaming exposed a few load-bearing assumptions with this approach.[2] First, and perhaps most fundamentally, DoD's approach conceives of requirements as gaps: shortfalls in capability, as predicted by future missions and projected adversary capabilities. This is explicitly represented in the JCIDS call for a Capabilities-Based Assessment (CBA). Second, there is an assumption that a formally recognized requirement (i.e., via JCIDS, an Urgent Operational Need, or some other means) must be established before resources are allocated to a

---

[2]   There are, of course, cases in which these assumptions do not hold. But they do hold in most cases, particularly for major capability acquisitions.

program of record. This assumption has been relaxed with the establishment of the Middle Tier of Acquisition pathway, but this remains an exception rather than a rule. Finally, there is an assumption that the requirement must be satisfied prior to fielding. As a capability evolves through the acquisition pipeline, the original requirement gains more specificity through, for example, CBAs, analyses of alternatives, technical specifications, T&E master plans, and contracts with performers. At any given level of resolution, those requirements represent an objective that must be satisfied for the capability to advance to the next step in the sequential process that culminates with a decision to field.

Our exploratory analysis and policy gaming suggest that these assumptions are misaligned with a Mosaic construct. In the technology and security environment envisioned with Mosaic Warfare, capability gaps will continually open and close according to countermeasure dynamics. In fact, in a highly dynamic threat environment, a gap might not exist for long enough or be sufficiently well understood to be formally closed through acquisition of a capability. More generally, relying on gaps to initiate capability development precludes forward-leaning acquisition approaches that capitalize on the emergent opportunities on which Mosaic Warfare seeks to capitalize. Finally, Mosaic Warfare hypothesizes that continuously fielding incremental upgrades has accumulating benefits, even if they do not fully satisfy the requirement or close the gap.[3] Under Mosaic Warfare, a solution that provides some operational advantage now and can be improved upon with time is preferable to a solution that fully fills the identified gap at some point in the future, when the gap might prove not to be operationally relevant.

Resourcing

The PPBE process, instituted nearly 60 years ago, is DoD's approach to explicitly linking resource allocation to strategic planning. PPBE seeks to accomplish this, in part, by drawing out planning, programming, and budgeting processes for DoD acquisitions. A general (and unsur-

---

[3] This latter assumption relates to the point made earlier in this chapter about the logic of fractionation mitigating requirements creep.

prising) finding from our analysis is that the PPBE and congressional appropriations processes have limited flexibility to accommodate a warfighting concept that relies on agility in terms of what capabilities are pursued.[4] Two features of the current PPBE process were identified by players to be particularly burdensome on acquisition under Mosaic Warfare.

First, under the current system, funding for a program must be requested approximately two years prior to the allocation of funds. This requires clairvoyance regarding future capability needs, a requirement that is particularly unrealistic under conditions of rapid technological change and a dynamic threat environment. In effect, in most cases, this budgeting approach forestalls the ability to rapidly allocate funding for a pressing capability need. It also results in capability development efforts that are based on a stale assessment of the threat environment.

Second, the primary means of reallocating funds in the year of execution—reprogramming—is inflexible. Reprogramming requires passing an arduous congressional approval process. Besides slowing the process of funds reallocation, the need for congressional approval requires DoD to expend scarce political capital. When players explored the source of funding for Mosaic tiles, they commented on the impracticality of relying on reprogramming for capabilities of relatively low cost; put bluntly, the bureaucratic costs of securing congressional support outweighed the limited value of each individual Mosaic tile, and the value of the Mosaic force as a whole would get lost in the small-scale transactions required under the status quo.

Technology Transition

Technology transition is a process of connecting the scientific and engineering communities that are developing technologies (e.g., at DARPA, the service research labs, FFRDCs, industry R&D houses) with the capability developers in acquisition programs. Commonly known as the "valley of death," technology transition has received much attention in the policymaking and analytic community (for one example,

---

[4]   Eric Lofgren of the blog *Acquisition Talk* has compiled an excellent list of calls for budget reform from prominent sources. See Lofgren, undated.

see Landree and Silberglitt, 2018). Today, the services and industry partners manage technology transfer in different ways, but, as a rule, technology transition is decentralized. Science and technology (S&T) organizations individually set up technology transfer offices to seek out partnerships, and programs maintain connectivity with their S&T bases, with or without support from headquarters.

Mosaic Warfare appears to require solving technology transition challenges at scale. Maintaining the ability to field operationally meaningful improvements year over year seemingly requires DoD to manage a pipeline of activity for capabilities that are in development, in production, and on the cusp of being fielded. This would require maintaining situational awareness over the relevant S&T base, continuously matching technological breakthroughs with transition partners in industry or individual program offices, and managing the long-term health of the S&T portfolio. Failure to maintain situational awareness over the relevant S&T base would result in delays or failure to exploit emergent opportunities. The historically mixed experience that DARPA, the Strategic Capabilities Office, and the service labs have with technology transition make it difficult to imagine that the relentless drumbeat of capability improvements envisioned under Mosaic Warfare could be sustained using a traditional decentralized technology transfer paradigm. One player summarized the issue succinctly: "For [Mosaic Warfare], you need a conveyor belt. It is difficult to maintain [situational awareness] over the pipeline, but it is very difficult to do a portfolio review, which would need to be a constant job."

Acquisition Oversight

The DAS has been subject to a seemingly continuous stream of reforms. Recent reforms include the Adaptive Acquisition Framework, which encourages matching of the capability being acquired with an appropriate acquisition pathway; a new Middle Tier of Acquisition pathway; new concepts for acquisition oversight necessitated by the recent delegation of acquisition authority to the services; and changes to DoDD 5000.01. Because of these reforms, commenting on the merits and drawbacks on the DAS relative to Mosaic Warfare is akin to hitting a moving target. Nonetheless, we can locate several fundamental issues within today's acquisition oversight apparatus that appear misaligned with Mosaic Warfare.

First, the central organizing element in the status quo system is the program. Program offices receive program funding, typically through annual budget line items, and are charged with developing, testing, and procuring technologies that meet preset requirements. One practical effect of organizing weapon systems acquisition around specific programs is that it can associate careers of programmers, program management, budget staff, and others with the fate of programs rather than that of the overall mission. In some cases, this can create an incentive to prioritize the program's continued existence, even if cancellation or restructuring might be a more desirable course of action in view of the program's overall contribution to mission effectiveness.

Second, the program-centric paradigm predisposes the use of program cost, program schedule, and program performance metrics to assess the health of a given acquisition. Mosaic Warfare entails decreasing the amount of time until fielding for a large and diverse pipeline of continuously evolving capabilities. Management of such a system requires metrics that embrace the acquisition system as a process without a defined end state.

Finally, in the current program-centric model, cost remains the central measure of risk and thus determines the degree of oversight that is afforded to a program. Programs that are large in dollar value, such as Acquisition Category (ACAT) I programs, are subject to more-stringent oversight than programs that are smaller in dollar value (e.g., ACAT III programs). Because of the potential for Mosaic Warfare to result, via fractionation, in smaller average program sizes, an oversight model that measures risk based on tile-level dollar values might not be appropriate.

Source Selection and Contracting

Source selection is the process of choosing a performer to develop a capability. Contracting is the process of putting in place the necessary legal agreements with a selected performer. Both source selection and contracting can be lengthy processes, with analyses of alternatives and contract negotiations that span many months, if not years, for some complex programs.

Today, the DAS provides various means to streamline the contracting process. The use of OTAs, which were pioneered at DARPA and have recently seen more-widespread adoption, has been particularly effective (Mayer et al., 2020). Although further expanding the use of OTAs warrants consideration, open questions remain about which contracting vehicles would be appropriate to sustain a pipeline at the scale that is likely to be required in a Mosaic construct. FAR requirements are intended to ensure that contracting processes are fair and that the government is well positioned to steward resources—waiving these safeguards at any scale introduces risk into the system.

Our gaming highlighted a few specific barriers associated with source selection and contracting approaches. First, several players noted the relatively quick contracting decisions that will be needed to deliver capabilities on the timelines envisioned for Mosaic Warfare. However, as one player noted, the very steps taken to speed up the process, such as concurrent work and accelerated milestones, are associated with risks that have been previously identified and are mitigated by the current rules. Changing those rules will easily be cast as adding risk. In particular, players raised concerns about the ability of losing bidders to delay progress. For example, a player noted that businesses can and do advocate to leadership when their proposals are not selected. Here, again, waiving existing processes to gain speed could be portrayed as increasing risk and might impose time delays that must be worked out with oversight entities.[5] Regardless of the specific avenue through which concerns are raised, the core issue was captured by a player who said, "My point is that there is a person [somewhere] who will believe that you skipped over requirements and that the overall purchases . . . might be acquisition malpractice."

Second, players pointed to the risks associated with managing an industrial base dominated by small players and a changing demand signal. As noted earlier, an acquisition pipeline populated with more-

---

[5] We note recent work by Arena et al., 2018, that suggests that delays are not as large as some have imagined.

heterogeneous and more-fractionated systems would likely create opportunities for more-nontraditional contractors. Although DoD has recently placed emphasis on gaining access to nontraditional suppliers via such defense innovation organizations as the Defense Innovation Unit, AFWERX, SOFWERX, and others, Mosaic Warfare will likely introduce a new challenge in which engaging smaller suppliers directly is a much more common occurrence. The advantages of potentially increased competition and innovation come with risks. One player noted, "I worry about the DIB [defense industrial base], because it is all little folks, but, in the background, we have lost the big providers, and now we have gotten rid of a lot of manufacturing practices." In addition to development and production capabilities, large defense-servicing firms have experience navigating the acquisition process and have relationships with key members of Congress. Because of lack of experience and relationships, new entrants might proceed through the acquisition system more slowly, at least in the near term.

Systems Engineering and Integration

Under the current system, systems engineering and system integration are large drivers of acquisition schedule, cost, and risk. However, Mosaic Warfare proposes a force characterized by less monolithic platforms and a high degree of composability. These features can be expected to defer a portion of systems engineering and integration tasks from the level of individual tiles to that of the Mosaic force package.

Test and Evaluation

Under the current system, T&E tends to be serial, and stages of the T&E regime correspond to stages of the acquisition process (e.g., passing Milestone C requires passing developmental testing and operational assessment). Reliance on remote or virtual testing is relatively low, and physical testing facilities are often heavily burdened.

DARPA has recognized a need for more-substantial levels of virtual testing given a Mosaic force construct. Because the theoretical number of force compositions scales exponentially with the number of tiles in the Mosaic force, it would be impossible to physically test the number of new force compositions introduced with each new tile.

It appears unlikely that virtual testing could fully supplant the need for physical testing. Individual tiles might need to be tested in operationally realistic settings, whether to verify system performance, test system integration, or ensure operator expertise.

Several barriers within the current system emerged during the game. First, players observed that Mosaic tiles would compete with other programs for access to scarce T&E infrastructure. Although DoD maintains a robust T&E infrastructure to support such testing, competition for scarce T&E resources is fierce—and the sheer number of capabilities that Mosaic Warfare aims to transition on an annual basis suggests that the aggregate demand for T&E infrastructure at the Mosaic level could easily outstrip supply. Although the infrastructure demands associated with virtual testing are lower than for physical testing, large-scale virtual testing still requires significant testing capacity and resources. Observing this reality, a player commented, "We do not have the virtual test network on a scale that would be able to work with this theoretical system of testing all components within a portfolio." The problem becomes even more acute for physical testing, particularly operationally realistic testing of many systems simultaneously. For example, players suggested the value of a designated exercise that would allow for integrated testing of systems, perhaps modeled on red flag and other high-end training exercises. However, even running such an exercise once a year was seen as a problematically large commitment of resources, while, at the same time, an annual cycle was too slow and inflexible to match the Mosaic vision. Relatedly, players noticed capability gaps in particular aspects of the DoD T&E system. Observing a lack of capacity to conduct cross-domain testing, one player stated, "One of the biggest challenges from a testing standpoint is that we lack the ability to test the products across multiple domains." Because the Mosaic concept places value on exactly this kind of cross-domain capability, current testing structures might not be able to provide a rigorous evaluation of key Mosaic capabilities.

The second barrier suggested by players was that the current T&E approach, which tends to treat threats as static, fails to test one of Mosaic Warfare's important value propositions: a focus on a dynamic, technologically sophisticated adversary. Describing this gap in the current T&E regime, one player noted, "This is a very clear shortfall of our system, and it's an issue of how the developmental testing lacks the ability to replicate the threats we are envisioning." This could be related to the broader issue of how progress toward fulfilling requirements is measured under the new system, since a dynamic adversary is part and parcel of the drive toward evolutionary measures. Here, again, the issue of developing metrics appropriate to understand evolutionary progress becomes critical to building a more robust system for Mosaic acquisition, because, once such metrics are in place, it will be easier to determine what tests are needed to measure progress.

Fielding and Sustainment

When it comes to fielding, a slow acquisition process might offer substantial benefits. For example, a slow cadence of capability introduction might provide time for the services and COCOMs to gain familiarity with the emerging capabilities; experiment with early prototypes; develop TTPs; train current and future generations of operators, maintainers, and logisticians; and integrate capabilities into war plans. A dramatic acceleration in fielding, which is envisioned in Mosaic Warfare, could erode these benefits and strain the capacity of the system to build the familiarity, expertise, and processes required to effectively field and sustain capabilities. It seems self-evident that the traditionally sequential processes of T&E, TTP development, training, and experimentation might need to merge to accommodate the increased throughput of capabilities under a Mosaic vision.

Summary

To the extent that Mosaic Warfare entails accelerating capability development and fielding as the rule rather than an exception, there appear to be barriers that are broad and systemic; large improvements in one acquisition function or marginal improvements in a swath of functions are unlikely to achieve DARPA's vision. Fully embracing Mosaic Warfare at scale is likely to require substantial shifts in the way DoD implements the management controls that are currently embodied in the requirements, resourcing, and acquisition systems.

CHAPTER FIVE

Insights About Mosaic Warfare Under an Alternative Acquisition Paradigm

In this chapter, we discuss a series of insights about acquiring a Mosaic force under an alternative vision for acquisition centered on the JMO described in Chapter Two. These insights are summarized in Table 5.1, which is provided at the end of the chapter. Evidence from the second and third activities (i.e., days 2 and 3 of the capstone game) suggests that the JMO model might have the ability to overcome some of the barriers in the current system. Across the areas of requirements, resourcing, technology transition, acquisition oversight, source selection and contracting, systems engineering and integration, T&E, and fielding and sustainment, players felt that the alternative system provided them with more flexibility. Game play also pointed toward particular insights regarding governance and Mosaic Warfare that are described in the final section of this chapter. However, the proposed alternative model was not a cure-all—new risks and challenges were also highlighted, suggesting the value of further analysis of organizational alternatives.

As in the previous chapter, we selectively draw quotes from anonymous players in our game as a way to illustrate certain points using language expressed by the players themselves; unless otherwise stated, we make no general claim as to whether the quotes represent an agreed-upon position of the group.

Requirements

With the proposed alternative model, we attempt to align the requirements process with Mosaic Warfare in a few ways. First, the mere establishment of an ASuW mission office—the ASuW JMO—represents a top-level articulation of a requirement. By establishing the office, the Secretary of Defense and the Chairman of the Joint Chiefs of Staff in effect communicate a priority for a particular mission in a particular theater. Although we did not explore the topic deeply, we envision a world in which a small number of JMOs could be set up, each representing a different mission and theater.[1] The Secretary of Defense and the Chairman of the Joint Chiefs of Staff could weigh in on priority requirements by empowering and disbanding an existing JMO or establishing new JMOs. By articulating requirements at the level of a mission rather than the traditional capability gap identified in a CBA, this approach leaves more space for opportunistic advances that might not be pre-identified to be prioritized for acquisition. It also focuses attention on holistic gains rather than narrower problems, increasing focus on how the force will fight as a whole.

Second, the model gives the JMO the authority to set needs on an annual basis within the scope of its mandated mission set and theater. As envisioned, these needs represent guidance to the development and acquisition communities on areas of attention, not an indelible objective that must be achieved. These needs would be informed by the MCC (i.e., via analysis). This provides focus, without requiring that a capability fully meet a preset goal to be deemed useful.

Third, the model ensures the services' awareness of needs and their ability to express objections to the so-called -ilities (e.g., maintainability, trainability, supportability). In our policy game, we allowed the JMO to set the requirements for life-cycle supportability (e.g., the expected service life of a given tile) and reliability (e.g., the minimum performance threshold) and allowed the services the ability to offer their input before proceeding with the acquisition. Although it is hard

---

[1]   Some players asked how many JMOs could be possible and highlighted an issue regarding seams between missions, which we discuss later.

to draw firm conclusions from just a few instances of game play, the adjudication of that condition forced the players to drive toward consensus regarding the system attributes of the tiles under consideration.

Finally, and most critically, the concept leaves the ultimate decision on fielding to a JMO-service-COCOM triumvirate that must reach concurrence. We did not specify the criteria that the group must use to make its decision but instead envisioned that its decisionmaking would be informed by its members' competing interests in current operations, a cost-benefit analysis, the sustainability of the force, the safety of operators, and the health of the S&T base, among other factors. This arrangement avoids the challenge of requiring a program to meet "prespecified requirements" criteria. Put simply, if a capability is seen as useful to the COCOM and does not create fielding and sustainment challenges that outweigh those benefits, it can be fielded.

It is worth noting that although affording veto power over fielding to the JMO and to the relevant service and COCOM might decrease institutional resistance to the JMO-centered model, it might also create a new bottleneck. For instance, requiring three-party concurrence prior to fielding might result in organizations wielding their veto power to advance organization-specific interests at the expense of acquisition throughput. Future research should consider how such a system might evolve over time and across multiple JMOs to ensure that these incentives do not generate negative behavior.

During game play, an additional question emerged: How can a JMO gain support within DoD and Congress without a requirement that serves as an agreed-upon benchmark for progress? One player characterized the issue as follows:

> One of the greatest challenges is that when you don't have clear requirements . . . there is no way to measure [your progress]. . . . You are vulnerable from attacks [from the institutional competitors] saying that you did not meet your goal. [In that case], the only way to know for sure is if combat breaks out and you see that the system or tool works as required and promised.

This comment highlighted the perhaps understated value in having a defined requirement; if nothing else, it provides a recognized benchmark for programmatic success shared across major stakeholders. In our policy gaming, the potential for a veto created an incentive for the JMO to respect the service and COCOM equities throughout the development process so that, by the time a capability was considered for fielding, there was alignment on what standard should be used and relatively little debate over the fielding decision.

Resourcing

In the JMO-centered model, we seek to ameliorate the resourcing bottleneck by providing the JMO with a two-year fully fungible mission fund—removing the need to anticipate future needs and adding flexibility that is currently lacking in the system. The JMO is stipulated to have full discretion over the use of this fund, allowing for the in-year allocation of funds to high-priority tasks and the rapid termination of low-priority efforts. In our model, the mission fund is approved by Congress as a single Program Element. We set the JMO budget at $2 billion, reasoning that this was large enough to make meaningful progress but perhaps small enough that Congress might be willing to treat the Program Element as an experiment. However, in practice, gaining congressional support for a large, flexible mission fund would require DoD to carefully articulate, and build support for, the argument for the fund.

In our second activity, players were presented with a portfolio of activities to prioritize given a fixed budget. This forced players to make judgments about how to fund emergent capabilities under a resource constraint. It is worth noting that the mere stipulation of a JMO mission fund effectively removed the resourcing bottleneck that provided challenges under the current system. That is, when players wanted to fund a development effort, they were able to do so during the year of execution. The two-year gap between need identification and resource allocation is simply not present when discretion over the allocation of funds is granted to an office responsible for determining what capabilities are needed. This is not to say that the alternative acquisition model proposed in the game solved the resourcing problem—the game took as a given that initial congressional approval could be secured rather than testing this point. Addressing DoD's resourcing problem might be largely a function of attaining political—and particularly congressional—support for greater discretion over the allocation of resources.

Technology Transition

Conceptually, the proposed alternative acquisition structure addresses the challenge of technology transition (i.e., the need for greater capability pipeline management anticipated by a Mosaic force) by giving the JMO the explicit responsibility of managing technology transition for the entire joint ASuW mission area. This elevates responsibility for technology transition to the mission level and creates a single point for technology developers to approach in finding transition partners in the services and elsewhere. The JMO also serves as a source of resources for services to sustain their own S&T pipelines or to accelerate transition for technologies on the verge of a breakthrough. Players agreed that the JMO as proposed had the potential to mitigate this technology transition challenge. As one player commented,

> I don't think you want the JMO owning the S&T, but I think you want to make it easier for the S&T to get into the Mosaic pipeline. The way the JMO was doing the development in this scenario was [one way]. Maybe this is how you get through the valley of death, because you don't need to worry about the life cycle, which would allow for them to cut service costs.

However, the game was not designed as a robust test of this aspect of the concept. In practice, there might be pitfalls associated with a more centralized approach to technology transfer. Obviously, assigning responsibility for technology transition to an organization like the JMO would not solve the problem by itself. For now, our research merely identifies technology transition as a key challenge to implementing Mosaic Warfare.

Acquisition Oversight

The JMO concept offered an approach to moving away from program-centric oversight by elevating acquisition oversight to the mission level. That is, the JMO is an organization charged with ensuring that the Joint Force is equipped to effectively execute a given mission set; it is not organized based on the particular programs it oversees.

Perhaps the principal benefit of orienting acquisition around missions is to correctly locate the role of weapon systems as means to an end.[2] The purpose of a military weapon system is to contribute to the realization of mission objectives. Under the current, program-based model, the instrumental nature of weapon systems can be lost. The use of a mission-central paradigm, as proposed in the JMO model, might focus incentives on mission outcomes and advance mission objectives by tying careers of programmers, program staff, and budget staff to mission execution rather than to program outcomes.

With the proposed alternative acquisition model, we also sought to eschew program-centric metrics in favor of mission-centric ones. The principal metric used in the game was *net mission effectiveness*, a measure of the marginal increase that a given capability adds to expected mission performance. For our purposes, we stipulated that this metric was generated by the MCC, which was assumed to be technically feasible and to produce bureaucratically acceptable analysis.

Our policy game revealed several open questions related to oversight—in particular, how to conduct oversight given the smaller programs entailed by Mosaic Warfare, how to measure acquisition performance in a Mosaic context, and how to oversee a large and diverse capability pipeline. The remainder of this section describes these questions and how they manifested during game play.

Program Size and Risk

Historically, level of oversight has often been driven by program size. The largest programs, ACAT I programs, receive the greatest oversight.

---

[2]   This reframing is not limited to Mosaic Warfare but is shared with other innovations in the acquisition community, including mission engineering and CBA.

Under the JMO concept, it is not clear how these standards would be applied. If tiles are treated as programs, they are likely to be classed as ACAT III programs and subject to less oversight. This status might appear advantageous in the sense of quickly advancing individual tiles from developing to fielding,[3] because, in the words of one player, "as the level of dollars increase, it creates antibodies against you." But at some point, less oversight or management controls will translate into more risk, which would compound if applied to potentially hundreds of Mosaic tiles that ostensibly would be in the pipeline at any time. How these risks aggregate is currently not well understood. One player observed that although the individual programs constituting the capability portfolio used in the game were ACAT III in terms of size, when combined, "the portfolio is ACAT I scale" and might require a greater degree of oversight. Somewhat separate from the need to objectively understand the risk of the JMO as a whole, there was also a concern that Congress might not allow for a meaningful decrease in overall oversight. One implication of these discussions is that novel means of measuring risk and determining oversight thresholds should be explored.

Appropriate Metrics for Mosaic Warfare

The policy game revealed a need to conceive meaningful measures of merit for Mosaic Warfare. Our initial research suggested that throughput (or time-effectiveness), which measures the marginal increase in mission effectiveness per unit of time, constitutes a reasonable measure of acquisition performance under Mosaic Warfare.

The game suggested the need to measure another core Mosaic value proposition: adaptability. Mosaic Warfare seeks to allow DoD to adapt to changes in external conditions on both force development and operational timelines. A measure of merit is thus needed to assess the robustness of Mosaic Warfare to uncertainty. Previous frameworks— such as robust decisionmaking (Lempert, Popper, and Bankes, 2003)

---

[3] A program's ACAT status is determined largely by the eventual total expenditure for research, development, and T&E or procurement. A program's ACAT status determines the decision authority for the program and thus the level of oversight provided.

and assumptions-based planning (Dewar et al., 1993)—that focus on assessing decisionmaking under conditions of deep uncertainty could help to first conceptualize and then measure the robustness of Mosaic Warfare to uncertainty. Metrics that compare Mosaic Warfare with the status quo system might reveal the advantages of the Mosaic concept and point toward means of increasing its robustness to uncertainty.

Game play suggested that the portfolio management–oriented metrics at the tile level proposed within the game were deficient. In the game, players were provided information about estimated cost, timeline to essential operational capability (EOC), and net mission effectiveness. Players debated how to use these in their decisionmaking—while some players gravitated to a single metric, others argued for balancing between them, either by using a balanced scorecard approach or by altering which factors are prioritized over time. That said, even an approach that integrated all three metrics was not broadly viewed as satisfactory. As one player put it, "[You] would need a whole other set of metrics . . . because the Mosaic vision argues that the composite of all the elements may be more than the sum of its parts." Another player worried that without a measure that could provide that context, "we don't have the judgment available to know what Mosaic is meant to be."

However, players did not have a clear vision of an alternative suite of measures that would better contextualize each element in the kill chain. For example, a couple of players argued that it would be helpful to see what role each element played in the kill chain to gain a sense of the portfolio's balance on this dimension (e.g., including a field that could specify whether a sensor would be appropriate to find or fix a target). However, this was seen as only a starting point. The end goal was more ambitious: to reflect the fungibility of systems across different chains and the synthetic gain in effects envisioned with Mosaic Warfare. One player expressed the broader need as follows: "I want effects webs that you can move rapidly within and want as many options as can be shown and balance the effects that these things can bring." While recognizing the utility of such a tool, players also noted the challenge of delivering complex information of this kind in a way that does not overwhelm a decisionmaker. Current work on analysis and visualizations to support mission engineering threads and capability portfolio management in traditional acquisition spaces might offer new approaches to this type of analysis.

Pipeline Management

Our engagement with the JMO concept suggests the need for a new oversight function: capability pipeline management. Mosaic Warfare entails a robust capability pipeline. At any given time, dozens, if not hundreds, of tiles will be at various stages of maturity. Such a pipeline involves managing myriad trade-offs between fielding a capability today, terminating an effort, investing in the future, incentivizing innovation and competition, and not overwhelming operators. With regard to capability pipeline management, players focused on two key trade-offs: the number of programs in the system over time and the mix of programs.

The first challenge articulated during game play was a need to balance quick wins with meaningful gains over time. Both the research team and the players expressed a tension from the desire to push programs that would mature and field quickly, to meet the Mosaic Warfare goal of rapid fielding, and the concern that such a strategy would neglect development of new capabilities that ensured long-term pipeline health. A particular concern was that maturing capabilities, even under the Mosaic concept, might not proceed quickly. One player noted, "I can see the value of getting something out fast to beat the naysayers. . . . It usually takes several years to make something worth transitioning." In the context of a multiyear development timeline, preserving the ability to offer future value dominated some players' thinking.

Independently, there was a concern that fielding many programs at once would overwhelm service and COCOM capacity to integrate new tools, preventing the warfighting force from gaining the full benefit of the investments. This concern created great focus on the need to manage the flow of programs over time. One player summarized it as follows:

> But a lot of the issue is that we have tension between racing to
> have faster things coming to EOC, but have you emptied the

cupboard, and do you have a healthy, long-term pipeline for this? I recognize that [DoD has] a limited capacity to assimilate all these things. . . . I need to strike a balance between the current programs and the new capabilities offered.

Although no consensus solution to the challenge of balancing quick wins with long-term impact emerged, players offered a few potential strategies worth further consideration. The first was to do as much as possible to smooth the pipeline over time for a consistent flow of new capabilities. This approach generated such decision heuristics as "shrink the number of programs coming off development into fielding so services are not overwhelmed." A second approach was to vary strategy over time. One example of this approach that was raised several times was to try "for as many capabilities to show up as early as possible on the first turn and then narrow for effectiveness." In other words, this strategy was to make many quick investments to see what panned out and then to develop more-tailored investments in the future to build on the successes. This approach to what we might call "intentional lumpiness" depends far less on getting an exquisite answer in advance than on rapid prototyping. It also implies that, year to year, the contributions to the Mosaic force will look very different, which raises interesting challenges about how such an approach will be perceived at various points in time. In game play, a smoother approach was eventually selected, but, given the robust discussion of the second strategy, it likely deserves further consideration.

The second challenge focused on program size rather than timing. There was a concern that larger programs offered larger gains but also reduced flexibility. One player captured the challenge as follows:

The challenge for how this is set up is that the more expensive projects have very high net effectiveness, so you are loath to take them out, but the smaller ones have a lot of flexibility and do interesting things.

In other words, some players were drawn to investing in a few big-ticket items based on relatively high promised effectiveness but believed that this strategy generated risk. Players also expressed concern that investing in fewer, highly effective programs pulled them away from the stated characteristics of Mosaic Warfare.

As with the challenge of balancing tiles over time, players eventually settled on a mixed strategy, but alternatives are worth further consideration. In particular, varying the provided mix of potential Mosaic tiles in future games might help improve our understanding of whether this dynamic was a relic of the specific mix of systems we developed for game play or a more generalized tension that must be grappled with under Mosaic Warfare.

Source Selection and Contracting

In the proposed JMO model, we attempt to address the potential barriers created by contracting and firm selection by establishing a preapproved, JMO-administered vendor pool that services and COCOMs can use to quickly gain access to performers. This feature of the alternative model attempts to hasten the end-to-end weapon acquisition process by accelerating time-consuming legacy processes. Legacy functions that can be anticipated to be largely obviated by the proposed vendor market include those associated with the market research obligation (FAR Part 10, 2021), such as assessing the state of commercial alternatives, finding related historical acquisition programs, and gathering price information. Legacy functions that can be anticipated to be accelerated by the proposed vendor pool include contracting and the traditional vendor solicitation, evaluation, and source-selection process.

Our policy game was not explicitly designed to explore the concept of a preapproved vendor pool. However, tellingly, at least one player endorsed the concept without being prompted. The player observed that it would be helpful to have multiple vehicles for multiaward contractors with a flexible ceiling, which would allow a contractor to move quickly, in a way that would be difficult with a competed contract.

During enterprise-level game play, we also leveraged the preapproved vendor pool concept by providing players with the "keep warm" option for a development project that players were unable to prioritize in a given year of execution but wanted to retain as an option to fund in future years. This option was valued as a way to ensure "a diversity of options" by partially funding a given development effort, but players raised doubts about the practicality of its implementation. For example, one player pointed to the potential for congressional resistance to a "keep warm" resourcing option, arguing, "Congress will get mad about it. You could keep a few warm, but if Congress is looking for money, they will find it." In the end, players believed that the degree of resistance would be tied to the overall acceptance of Mosaic Warfare.

Systems Engineering and Integration

As discussed earlier, Mosaic Warfare defers systems engineering and integration from the level of tiles to the level of the Mosaic force package, ostensibly simplifying the systems engineering challenge that usually manifests at the program level. Our study has made the generous assumption that technical interoperability is seamless. But multiple players highlighted a challenge that cannot be assumed away—physical integration of Mosaic enablers into existing platforms. To the extent that the Mosaic model envisions hosting, for example, Mosaic munitions, EW payloads, and communications gear on traditional platforms, potentially including capital ships and combat aircraft, there might be a challenge in doing so in a way that accommodates the size, weight, power, and other constraints associated with those platforms. One player offered the following framework for thinking about this variety of integration from the service perspective:

> I [am thinking about] the rubric of do no harm. . . . Does [a Mosaic capability] do harm to the platform that already exists, and how do we figure it out through the testing regime? That's easy with an aerostat, but if you're putting it on a ship, does that mean you need to dry-dock the ship to install, and does it cause malfunctions . . . ? I don't know how you do this within the existing system to test all the additions. It also defeats the speed idea in the system we are designing here. I am concerned about the interoperability question that you won't find problems until the

item is installed. For ships, dry-docking to install or if there are issues also creates readiness problems.

Several key ideas emerge here. The first is the difficulty of predicting or identifying potential problems through existing approaches to testing (discussed in more detail in the next section). The second is that the need to complete complicated installations requires time, both delaying Mosaic acquisition and taking a traditional asset out of service while the work is being done. Each of these issues increases the risk, and thus the potential resistance to acquiring the tile in the first place. This reality is less a barrier within the current system than a reality of physics. The alternative model did not explicitly accommodate this in the proposed governance structure, except to allow the services and COCOMs an ability to veto a capability before it is delivered if the capability does not meet its constraints. Presumably, these criteria could be folded into the service decision space.

Test and Evaluation

Although the vision of the JMO offered additional resources for virtual and mission area testing to solve some of the capacity challenges identified under the current system, several unanswered questions remain regarding how T&E might appropriately assess the Mosaic force rather than a single capability.

First, the game raised the question, "What is the appropriate baseline for T&E under Mosaic Warfare?" One player articulated this point as follows: "What you have here is a very platform-centric testing process. You can think of this like a dev-ops system where I can plug it in and see where it fits in the Mosaic." In other words, traditional testing is about how a single system works, perhaps integrated with a small handful of systems. Because the value of Mosaic Warfare is predicated on the complex as a whole, testing must look at the whole in a way that it rarely does today.

A related point is that, because of the need to understand the value of tiles as the Mosaic force shifts over time, the traditional line between testing and implementation might no longer apply. In the words of one player, the shift to Mosaic Warfare requires a shift to "a model of continuous integration, as counter to the industrial model, and moving to an assessment model where testing and implementation are not separated. What is the model that allows for development and assessment as a process?" Put differently, because tiles are being continually updated individually, because the suite of tiles in the Mosaic force is being refreshed regularly, and because the threat that the Mosaic force aims to counter changes as adversaries update their strategies, test results can no longer be viewed as a single, static check at the transition point between development and fielding. A new vision that allows for these more dynamic changes would be needed, beyond what was envisioned in the current iteration of the acquisition game.

Three additional potential complications that such a new system of T&E would need to grapple with were legal requirements, funding, and tactics generation. First, one player raised a concern about the compatibility of the current, legally required OT&E and developmental T&E processes and the rapid fielding described during the game. As Mosaic testing moves away from the traditional model, these legal requirements might need to be adjusted to match.

Second, in the current paradigm, programs typically pay for T&E. As a result, T&E competes for other program-allocated resources, including upgrades and procurement. The game experimented with providing a T&E resourcing pool to the JMO so that these allocations could be made at the mission level. Because of the potential for T&E to create a bottleneck in fielding, this solution, along with other policy options, should be explored further.

Finally, a player noted that, under a Mosaic Warfare construct, T&E will have to incorporate the tactics-generation process:

> [T]actics and training is never talked about in the acquisition world, and, in my view, to get closer to Mosaic is how to get OT&E to bleed over into training and tactics generation, and that is logical and happens at a small scale all the time.

In other words, because so much of the value of a specific Mosaic tile is in how it is composed with other tiles as part of a tactical concept of employment, to have a valid test of the utility of a system, the process of tactics generation will need to be part of testing in a way that it is not currently.

Fielding and Sustainment

In contrast to the current slow cadence of capability introduction that allows for the force to accommodate new kit, Mosaic Warfare envisions a seemingly continuous stream of new or upgraded capabilities hitting the force. The risk was neatly summed up by one player as follows: "From the service perspective, you're changing too many ideas midstream. There's a lot to like here from the services, but you're constantly changing stuff off and you're in constant reassess and training. Too much churn in the system." Managing this churn will present unique challenges for the nonmaterial capability developers who develop doctrine in the traditional paradigm, planners who need to make informed judgments on whether to embrace Mosaic Warfare for specific missions, operators who might be asked to entrust their lives to the new systems, and political and military leadership who will require or provide professional military judgment about the utility of Mosaic Warfare.

These points were not lost on players in our policy game. Players stressed that unless Mosaic systems were carefully integrated into the force, they would fail to provide value. Two key barriers were identified. First, operational forces need to understand how new capabilities could contribute to missions. One player suggested that "there is an operational component that goes along with this that needs to see if the warfighters will actually adopt the technology or capabilities developed." Without the concepts and TTPs that operationalize a system, adoption is unlikely. Second, commanders need to be confident that, during operations, new capabilities will perform as advertised. One player framed the issue as follows: "From a service chief perspective, I really don't want to trust a system until we see it demonstrated by real users in an operational environment." If there is not enough time for operational testing and use in low-risk settings, commanders' trust might not be sufficient to use a system during conflict.

The alternative acquisition model that we propose acknowledges the services' finite capacity to absorb new capabilities by giving the services and COCOMs the ability to veto or pause any capability that the JMO presents for fielding. In effect, this would allow the institutional players to "hit the brake" when the rate of fielding exceeded their capacity to absorb. Because of the likely pressure on the services and COCOMs to field ready capabilities, it will be important within this context to take measures to ensure that a temporary pause is considered an appropriate action meant to assure efficient fielding. Another self-limiting factor would be resource scarcity. To manage a balanced pipeline, the JMO would be required to strike a balance between investments that field forces today and investments that yield new capabilities in the months and years to come. The resolution of our policy game was insufficient to assess whether these measures would be sufficient and what side effects (most obviously, delays) these tools might produce, but, for now, we highlight the issue.

Also related to fielding, the question of how decisions to field incremental upgrades would affect adversary decisionmaking surfaced during our policy gaming. This was on display when players were presented with the decision of whether to field a capability that had reached maturity, and the services, COCOM, and JMO were debating the merits. According to one player,

> What was postulated is that the ELINT sensor is a single response to the [adversary] sensor they just launched, and then the [adversary] will just move on after ours is revealed. So, our issue is that if we reveal, we make our new development useless, or do we not deploy the item to maintain counter as a classified program? Answering this question is hard because the development might be too tied to the adversary's decision. When you mix integration and testing, it becomes more expensive and time-consuming.

Although DoD institutionally is very experienced with reveal-conceal decisionmaking, Mosaic Warfare raises the question of whether, or in what circumstances, marginal upgrades would be subject to special access–type security restrictions. If the Mosaic force is as diverse as imagined, it might be inherently resilient to individual adversary countermeasures. Furthermore, if upgrades are as continuous as imagined, the demands for secrecy might be mitigated if the next upgrade down the proverbial pike has the potential to render the current one out of date.

Governance

We conclude this section with broader discussion of the JMO concept from a governance perspective.

First, and unsurprisingly, our policy gaming confirmed that Mosaic Warfare is sufficiently radical that it will elicit antibodies from existing DoD organizations (represented by the institutional team during game play). And for good reason: The services and COCOMs have a legitimate need to protect their equities under Title 10, and any proposed governance alternative must embrace this. One player, during the day 1 activity that focused on assessing the legacy acquisition paradigm, commented, "You need a single point of accountability for the Mosaic," acknowledging that DoD would resist Mosaic Warfare unless there was a top-down effort to institutionalize it. Another underscored the inherent jointness of Mosaic Warfare, anticipating the jointness of the JMO: "The Mosaic concept works best when it works across multiple services." These remarks foreshadowed the concept of a JMO that reported to the Secretary of Defense, reinforcing the value of that aspect of the proposed governance structure.

Second, our policy gaming suggested that our proposed alternative could work by creating an incentive for compromise. The proposed governance structure gives the JMO authority and resources to move quickly while giving the services and COCOMs authority to slow down the process in light of specific concerns. Several players noted the need and the opportunity for compromise within the alternative system. In the words of one player, "The JMO has to be careful not to be seen as 'I am the JMO, and I'm here to help.' It needs a persona that works well with the services, and it needs to have good communication so there isn't friction and competition." However, it was unclear how these dynamics might shape the system over time. For example, over time, the JMO might find that its influence wanes and the system reverts to behavior more like that of the traditional acquisition system, losing some of the envisaged value of Mosaic Warfare.

Other players noted the need for the JMO to understand service equities and perception of risks to put forward proposals that work for both sides. This was raised both at the level of specific tiles (particularly those that would integrate with platforms operated by a service branch) and at the enterprise level, where a JMO that understood service priorities might be able to make smarter investments. The resources afforded to the JMO created an added incentive for the institutional players to compromise because the JMO was seen as a potential bill-payer. In multiple instances in which players representing the services and the COCOM were confronted with JMO decisions that posed modest challenges for their core interests, the institutional players fell back on the realization that they were not paying, giving in to the JMO's prerogative with the sentiment, "Well, the JMO is paying."

Third, our policy gaming suggested that the JMO might have a natural value alignment with the COCOMs. Although the JMO has a broader interest of maintaining a capability pipeline, institutionally its prerogative during the game was to increase the rate at which new capabilities entered the force to improve short-term operational effectiveness. This aligned well with the COCOM interest in deploying the best capability for the "fight tonight." Over the course of the game, perhaps because of this alignment in interests, the COCOM and the JMO appeared to find it easy to reach compromise solutions.

Fourth, the JMO introduced seams that were not fully explored in the game but that would need to be managed in practice. For example, in the proposed vision, the services maintain responsibility for major platforms, while the JMO has greater responsibility for Mosaic enablers. In the game move that focused on managing enterprise-level investments, players were presented with two investment options that were intentionally designed to test the boundaries of Mosaic acquisition: a flotilla that served as a missile barge, challenging USMC equities, and a single-use bomber, challenging USAF equities. Both platforms were designed to be attractive investments from the perspective of their contribution to mission effectiveness. But given the division of responsibility for platforms and enablers, players hesitated. Illustrative of this hesitation, one JMO player commented, "The big nut to crack is the flotilla. It is reasonably effective, but it's borderline a platform, so we do not want to get into that." Another JMO player expressed a similar hesitation, commenting, "Single-use bomber is the best bang-for-buck, but I am principally opposed to funding it, as it is too much a USAF thing and doesn't fit into the island of misfit toys of Mosaic systems." In other words, when confronted with capabilities that exhibited platform-like characteristics, players felt that they properly belonged in the realm of the services rather than that of the JMO. However, because the game did not represent any service's full portfolio, we did not test whether the service would, in fact, have opted to invest in these small systems. As a result, there is a potential that such capacities might not have been acquired at all, even in the face of broad agreement about their value, because their profile made them neither fish nor fowl.

Another seam that was revealed during game play was the one between capabilities fielded by the JMO for ASuW and those that might also bear on other mission sets. This seam is likely to become even more evident with the introduction of new JMOs with new mission sets. For example, players noted that provided metrics related only to the value of a system in the ASuW mission, which would discount the value of versatile systems that could contribute to many mission areas. Establishing a means of measurement and coordinating effort and responsibility for tiles with application across mission sets remains a pending point of inquiry. Again, ongoing efforts to design analytics to support mission analysis might provide options to bridge the seam.

Fifth, players noted the challenge of transitioning to a Mosaic construct. Specifically, concern was raised regarding the initial rollout of Mosaic force elements. One player articulated this apprehension succinctly, noting, "Once you have a Mosaic system in the field, you can snowball off of it, but getting one started and successful is the biggest issue."

Finally, game play suggested an emergent governance question: "Will multiple AIs result in the playing of politics over which AI to use?" As AI-based decision aids proliferate across DoD, it is likely that distinct organizations will have their own preferred AIs, mirroring current debates over scenario selection. Therefore, for any given decision that involves the interests of multiple organizations, there is a possibility of competing AI-informed recommendations. During a dry run of the game, a player representing USN cited its AI as a potential competitor to the JMO's AI, the MCC. Further exploration of how to coordinate and deconflict competing AI is warranted.

Table 5.1 summarizes the identified barriers to acquiring a Mosaic force in the current system and in an alternative system.

Table 5.1
Summary of Issues and Barriers to Acquiring a Mosaic Force

Function		Issue or Barrier
Requirements	√	Presumption that a requirement must precede resourcing
	√	Presumption of a static requirement that must be satisfied prior to fielding
	❏	Can Mosaic Warfare build support without a requirement to serve as an agreed-upon benchmark for progress?
Resourcing	√	Two-year budgeting cycle requires clairvoyance on future needs, limits flexibility to adapt
	√	Reprogramming takes time, expends political capital
Technology transition	√	Difficulty in maintaining enterprise-view of technology pipeline
Acquisition oversight	√	Program-centric paradigm predisposes focus on program cost, schedule, performance
	√	Ten ACAT III tiles might equal an ACAT I in terms of expenditure yet receive less oversight
	❏	Need for a compelling measure of merit to guide Mosaic oversight
	❏	How can adaptability to the unknown (a Mosaic Warfare value proposition) be measured?
Source selection and contracting	√	Quick decisions risk protests from losing bidders, adding delay
	√	Increased throughput in contract actions will add delay
	❏	Risk in managing industrial base dominated by small players, changing demand signals
	❏	Players saw value in an explicit "keep warm" (to partially fund) contracting option
Systems engineering and integration	X	Mosaic Warfare defers integration risks from tile to Mosaic force package (assumed away in game)
T&E	√	Mosaic Warfare would compete with regular programs for scarce T&E infrastructure
	√	Programs pay for T&E; T&E competes for resources with upgrades, procurement, etc.
	√	Static threats would preclude testing one of Mosaic's value propositions (adaptability)
	❏	What is the baseline for T&E, given myriad potential kill chains?
	❏	How can legal requirement and timelines for OT&E be squared with Mosaic timelines?
	❏	Does T&E need to integrate with training and TTP development? And with requirements?

Table 5.1—Continued

Function		Issue or Barrier
Fielding and sustainment	X	Limited capacity of services and COCOMs to uptake new capabilities
	X	Sequential nature of requirements, T&E, and TTP development
	❑	Services, COCOMs, and JMO were able to reach consensus on fielding
	❑	Mosaic Warfare elevates the importance of near-continuous reveal-conceal decisionmaking
	❑	Risk that heterogeneity at the mission level scales unsustainably to chaos at the global level
Governance	√	The institution will fight back; need to protect enduring DoD equities and interests
	√	Mosaic value proposition is greatest in a joint context, but a joint context is hardest to implement
	❑	Will AI play politics? Service representatives might appeal to their own competing AI
	❑	JMO introduces seams between service and JMO responsibilities (e.g., enablers versus platforms)
	❑	Transitioning to a Mosaic force might be as hard as maintaining one in steady state
	❑	What is the endgame for JMO; should the JMO sunset?
New functions	❑	Pipeline management
	❑	Continuous Mosaic testing

NOTE: A green check indicates an issue with the existing system that is addressed by the JMO model. A red X indicates an issue with the existing system that is *unaddressed* or *reinforced* by the JMO model. A checkbox indicates a new issue or question raised by the game.

CHAPTER SIX

Alternative Models Beyond the Joint Mission Office

The process of experimenting with the JMO-centered model exposed a trade space of potential alternative governance models and management systems. In this chapter, we briefly discuss a few of the ideas that emerged during the exercise. Many alternative models, beyond those noted here, are likely to exist.

Mosaic Warfare as a Sustainment Activity

During a dry run of the game, a player surfaced the idea of instituting Mosaic Warfare as a single program and treating sequential upgrades to the Mosaic force as sustainment activities. That is, development of novel Mosaic enablers could be treated as a sustainment activity for an existing Mosaic-enabled program. This would allow the initiation of new development activity using the parent program's sustainment funding, thus allowing rapid resourcing of new Mosaic tiles. In this model, building a Mosaic force would, in principle, fall under current familiar models of governance and acquisition. That is, little legal or policy change would be necessary. However, a question for further research is whether this approach scales to all but the most-limited mission sets. We note that this option might be attractive as a starting point for Mosaic Warfare to gain a foothold in DoD, because it might be the easiest way for Mosaic Warfare to be "plugged in" to the existing paradigm.

Embrace the Middle Tier of Acquisition Pathway

Other players noted the recent creation of the Middle Tier of Acquisition pathway, which exempts service activity from traditional acquisition oversight within certain constraints. However, Middle Tier is still a relatively new acquisition approach. Over time, DoD will gain the data through experience to illuminate the costs, benefits, and risks of this approach as they pertain to acquiring a Mosaic force.

A Facilitator Model

Multiple players commented that the JMO model evoked too much of a "rapid acquisition cell," and they suggested an alternative "facilitator" model. Such a model might still instantiate a high-level organization like the JMO, but it would instead charter the organization to serve as a bridge between communities and to supplement programs with resources from a dedicated funding pool. (The new entity would lack many of authorities afforded to the JMO; for example, it would not publish needs; have direct oversight of service programs; oversee a preapproved contractor pool; or manage a special T&E resource). Because of its reduced infringement on existing institutional prerogatives, this alternative might be more palatable to a skeptical institution. A question for further research is whether a facilitator model adequately resolves many of the practical barriers that we discovered with the existing system.

The Alternative Acquisition Model Trade Space

Overall, we do not foresee a single best solution; our analysis foreshadows options and trade-offs between enabling Mosaic Warfare, balancing service and COCOM equities under Title 10, and minimizing the actual and political costs of the institutional changes necessary to realize a given option. Figure 6.1 illustrates a stylized conception of the trade space. At one extreme (the green dot toward the bottom left-hand

corner of the axes), using the existing system requires little or no implementation cost. However, our exploratory analysis finds this option to
be incompatible with scaling a Mosaic force. At the other extreme, the
JMO model is associated with the highest implementation costs among
the options we considered but provisionally has the greatest potential
to embrace the potential of Mosaic Warfare. Because we have yet to
fully explore the implementation costs and benefits of these models
vis-à-vis Mosaic Warfare, we are currently unable to locate the other
options in this trade space.

Figure 6.1
Alternative Acquisition Models Subject to Trade-Offs

Benefits (↑)   ?   ?   ?

	Status quo	Sustainment model	Embrace Middle Tier	Facilitator model	New JMO
Concept	Treat Mosaic tiles as distinct programs subject to JCIDS, PPBE, and Adaptive Acquisition Environment	Field an initial Mosaic as an ACAT II or III program under existing JCIDS requirement; subsequent tiles/ upgrades managed under sustainment program	Develop tiles through the Middle Tier of Acquisition pathway	JMO serves as a change agent, connecting developers, operators, and others	Consolidate authority for requirements, resourcing, and acquisition in new JMOs; limit responsibility to a specific mission, theater, capability (enablers), and forces to permit effective oversight
Pros	No cost to implement	Uses existing acquisition apparatus	Seemingly minimal cost to implement	Seemingly minimal cost to implement	JMO likely empowered to make Mosaic successful
Cons	Appears inherently incompatible with Mosaic Warfare	Need to tie to existing requirement, likely limiting scope Needs to broaden concept of a "program" to system of systems	Middle Tier not envisioned to handle end-to-end acquisition, including life-cycle sustainment; Middle Tier largely untested, future uncertain	Does it overcome identified functional barriers with status quo?	Unclear whether institutional equities can be protected if compatible with Title 10 Most costly to implement

Cost and risks of implementation (→)

CHAPTER SEVEN

Conclusions and Next Steps

This report has explored whether DoD's current governance structures and management systems for setting requirements, allocating resources, and acquiring weapon systems are compatible with DARPA's vision for Mosaic Warfare and, if not, what are viable options to improve or realign them in view of the vision. Our research is by nature exploratory, so definitive conclusions and strong recommendations are unwarranted. Nonetheless, we have reached several broad conclusions and discovered open questions, and we can propose several next steps for DoD and DARPA to consider.

Conclusions

On the one hand, we found nothing inherent to DoD's existing requirements, resourcing, and acquisition systems that prohibits or even inhibits the development, procurement, operation, or sustainment of forces that are more fractionated, more heterogeneous, and more composable. DoD's management systems and governance structures are by nature technology- and capability-agnostic and tailorable.

On the other hand, we found that the existing systems likely do not align with the Mosaic vision of fielding many new capabilities on operational timescales. Even if DoD could cost-effectively field systems that are more heterogeneous, more fractionated, or more composable, the amount of time required to execute the associated acquisition functions is likely incompatible with developing and fielding on the timescales envisioned by Mosaic Warfare. There are, in fact, ways of accelerating all of the functions inherent to fielding—from contracting to resourcing to T&E—but the options available to accelerate today are exceptions to a standard process rather than the rule. To the extent that DARPA envisions exercising those or similar options more routinely, there appear to be barriers that are broad and systemic; large improvements in one acquisition function or marginal improvements in a swath of functions are unlikely to achieve DARPA's vision. In our judgment, fully embracing Mosaic Warfare likely will require substantial shifts in the way DoD implements the management functions currently embodied by JCIDS, PPBE, and the DAS.

We explored an alternative model that was characterized by centralization of today's authorities for requirements, resourcing, and acquisition oversight within a new JMO reporting to the Secretary of Defense. Consolidating these authorities allowed the office to be decisive in taking action that would allow Mosaic Warfare to scale. However, our proposed alternative also limited the scope of the JMO to a specific mission, theater, forces, capabilities, and budget as a means of protecting service and COCOM equities under Title 10 and of promoting effective oversight. The proposed concept also featured a governance concept that required the joint approval of the JMO, the relevant service, and the COCOM before any new capability was fielded, effectively giving veto power to all three parties. We make no upfront claims about the optimality of our concept because this was the first phase in our iterative gaming approach to exploring Mosaic acquisition.

Through gaming and analysis, we found that this model could, in fact, mitigate many of the barriers that we discovered in the current model. We also found that it provided incentives for the players representing services, COCOMs, OSD, and the JMO to resolve competing interests between current operations, safety of operators, the cost and risks of sustainment, and long-term strategic and modernization objectives. Overall, we judge this alternative to be more likely to be effective and scalable in fielding a Mosaic force than is the current model, but it also creates new tensions with service and COCOM equities. More experimentation is needed to fully understand the first- and second-order implications of this concept.

Exploration of this specific alternative exposed a potentially vast space of other possible governance models and management systems. One proposed option was instituting Mosaic Warfare as a single program and conceptualizing sequential upgrades as sustainment activities. In this model, building a Mosaic force would, in principle, fall under current familiar models of governance and acquisition, but a question for further research is whether this approach is scalable beyond the most-limited mission sets. Some players commented that the JMO model evoked a "rapid acquisition cell," suggesting a "facilitator" model in which a high-level organization like the JMO (but without many of the authorities given to the JMO) serves as a bridge between communities and supplements programs with resources from a dedicated resource pool. This alternative might be more palatable to a skeptical institution. A question for further research is whether this option adequately resolves many of the barriers that we discovered with the existing system (a rapid acquisition mindset might be necessary even if it is insufficient for Mosaic Warfare). Overall, we do not foresee a single best solution. Rather, our work foreshadows options and trade-offs.

Next Steps

These conclusions lead to a few suggested next steps. First, we suggest that DARPA continue to experiment with alternative governance systems and management systems. The policy game that we designed proved to be a useful way of experimenting with alternative governance models. Placing DoD representatives in the mode of decisionmakers operating within the Mosaic model allowed for insights that would not have been easily deduced through mere logic. Although not reported here, various lessons were learned that could improve subsequent games—thus continuing the iterative learning process of formulating refined hypotheses, adapting and executing games, and so on.

Second, we advise that DoD embrace the following principles when considering alternative governance models and management systems:

- **Acknowledge enduring DoD needs for management controls for risk management and resource allocation.** Whatever barriers might exist within today's system, JCIDS, PPBE, and DAS allow DoD to manage risks and responsibly execute hundreds of billions of dollars each year. The fundamental need for risk management and resource allocation will endure even under Mosaic Warfare.
- **Acknowledge service and COCOM equities via Title 10.** Our analysis did not focus on the compatibility between Title 10 and Mosaic Warfare, but we found nothing to suggest any incompatibility. However, any proposed governance model must fully consider service and COCOM equities in legislation.
- **Embrace mission centrality in requirements, resourcing, and acquisition.** An important theme that recurred among players and the research team was the need to elevate oversight and measures of effectiveness above the program to the mission level. We did not arrive at a final conclusion. However, there appears to be wide consensus among participants about the importance of making mission centrality a focus of Mosaic Warfare. This aligns well with recent USD(A&S) and USD(R&E) initiatives to bring a mission engineering–based capability portfolio-management lens to acquisition oversight.
- **Embrace throughput (time-effectiveness) as a Mosaic Warfare measure of merit.** The Mosaic Warfare priority of decreasing the amount of time until fielding envisions Mosaic Warfare as a pipeline of continuously evolving capabilities. Management of such a system requires metrics that embrace the acquisition system as a process without a defined end state. Thus, such a metric as throughput—or time-effectiveness—to reflect the marginal increases in mission effectiveness per unit of time seems an appropriate alternative to program cost, schedule, and performance.
- **Define a measure of merit for Mosaic Warfare that embraces uncertainty.** A core Mosaic Warfare value proposition is adaptability—allowing DoD to adapt on both force development and operational timelines. A measure of merit is needed that cap-

tures the robustness of Mosaic Warfare to uncertainty. A body of prior work on robust decisionmaking (Lempert, Popper, and Bankes, 2003) and assumptions-based planning (Dewar et al., 1993) could help conceptualize and measure the robustness of Mosaic Warfare to uncertainty.

- **Develop a simulation of the Mosaic pipeline and use it to identify policy levers and bottlenecks that would inhibit realization of a Mosaic force.** Our work employed a policy game to examine issues and trade-offs. But it highlighted the important role of managing a capability pipeline with trade-offs between fielding a capability today, investing in the future, incentivizing innovation and competition, not overwhelming operators, and other priorities. Such a pipeline is amenable to modeling and simulation of Mosaic tiles through an acquisition pipeline. The model appears necessary to understand bottlenecks given the potentially large number of small tiles that are wending their way from the laboratory to the field.

Closing Thoughts

DARPA's vision of Mosaic Warfare is ambitious, compelling, and seemingly responsive to many attributes of the emerging technological and security environment. Transitioning this vision to widespread DoD acceptance might well require strong proponents across DoD to create change within institutions that today might—given their accrued equity in longstanding governance structures—often view the status quo as an end rather than a means. We advise the proponents of Mosaic Warfare to be mindful of falling into the same trap by making Mosaic Warfare an end rather than a means. Like all emerging visions for the future of U.S. warfighting, the ultimate test for Mosaic Warfare will be whether it allows the United States to deter and defeat adversary aggression.

Mosaic Warfare Scenario and Vignettes

In this appendix, we describe a set of vignettes that are representative of circumstances that DoD would confront when acquiring capabilities in a Mosaic Warfare construct. Explicit mention of *Mosaic Warfare* has been omitted by design to avoid unduly biasing results with player preconceptions of Mosaic Warfare as a warfighting concept; ideally, the players could arrive at the game without preconceptions. Per concurrence with the sponsor, all of the vignettes are situated in the context of ASuW missions in the Western Pacific.

The vignettes describe successful instances of Mosaic acquisition. They do not reference any specific acquisition governance model or management system. Instead, the narratives put forth generic descriptions of requirements, resourcing, vendor selection, T&E, fielding, maintenance, and sustainment events. This allows for the imposition of distinct acquisition systems over the narrative to assess relative performance.

During day 1 of the tabletop exercise, players were asked to identify the assumptions that would have to hold for the vignette to play within today's acquisition system. Players were then asked to assess how reasonable they believed those assumptions to be. Besides giving players a tangible experience of the difficulty of making Mosaic acquisition work in the current system, this exercise improved our understanding of the barriers to acquiring a Mosaic force within today's system.

During days 2 and 3 of the tabletop exercise, the vignettes were considered within the alternative, JMO-centered acquisition system described in Chapter Two. Holding the vignettes constant across the current and alternative systems allowed for a more straightforward comparison of system performance. The following sections first present the common force planning scenario in which all of the vignettes are embedded and then present the five vignettes, three at the capability level and two at the enterprise level.

Scenario

All of the vignettes share a common force planning scenario and were designed to provide operational and institutional context for the players. The scenario is summarized in Figure A.1 and in the following narrative:

> In 2030, the United States is committed to the principles outlined in the 2018 National Defense Strategy; military competition between the U.S. and China is contested and recognized. The United States has advanced a new Joint warfighting concept and fielded advanced capabilities to underwrite it but remains largely committed to a post–Cold War force structure & posture. However, the United States is at operational parity with China, with neither side confident that they could prevail in overt conflict.
>
> In order to regain overmatch, DARPA, in collaboration with the USAF and USN R&D communities, has demonstrated an emerging ASuW Mosaic. In three successive live demos in the South China Sea, the U.S. successfully demonstrates novel ASuW kill chains:
>
> - Demo #1: Attack submarines launch LR [long-range] Mk-48 torpedo with a glider swarm payload
> - Demo #2: A Small satellite cues an XLUUV to launch kamikaze UAV/UUV swarm
> - Demo #3: Novel ELINT sensor is integrated into aerostat and cues land-based SEAD [Suppression of Enemy Air Defenses] weapon against naval SAMs [surface-to-air missiles].

Each demo exercises a novel kill chain and visibly counters a Chinese capability. The sensors, munitions, and payloads associated with each demo are fielded within 6 months. The Secretary of Defense and Congress note success and move to institutionalize Mosaic Warfare by developing a new governance model and management system they believe will better support these emerging capabilities.

The core of the new governance model and management system developed by the Secretary is an ASuW Joint Mission Office (JMO) responsible for, and empowered to, equip the Joint Force to execute the ASuW mission set in the Western Pacific. The JMO is afforded significant authorities and resources for requirement-setting, resourcing, and acquisition. To allow effective oversight and to protect the legitimate institutional interests of the Services and COCOMs, the scope of the JMO is limited to a specific mission (ASuW), theater (INDOPACOM), weapon system type (enablers) and assigned forces. The JMO, Services and COCOMs must jointly approve anything before it is fielded. The sections to follow detail these authorities and elaborate the rationale for their inclusion in the model.

Figure A.1
Force-Planning Scenario Within an Acquisition Scenario

Acquisition scenario
- 2028 to 2032
- Strategic continuity (DoD committed to priorities of 2018 National Defense Strategy)
- Overall military competition between United States and China is contested
- United States has advanced new joint warfighting concept but remains committed to a post–Cold War force structure
- DARPA, in collaboration with USAF and USN R&D, demonstrates initial ASuW Mosaic
- Secretary of Defense and Congress note success, move to institutionalize Mosaic Warfare

Force-planning scenario
- 2035
- Chinese invasion of Taiwan
- Mission: ASuW

Vignettes

The first three vignettes presented a narrative of individual capabilities (Mosaic "tiles") making their way from development to fielding within a broader and more dynamic acquisition pipeline. These three capability "thread" vignettes are interwoven in a narrative that occurs over a period of three years (2030 to 2032):

- Vignette (Capability Thread) A—Leap Forward Sensor Tech
- Vignette (Capability Thread) B—Emergent Critical Requirement Gap
- Vignette (Capability Thread) C—New Force Multiplier.

Figure A.2 presents the relationship between the vignettes over a three-year period.

Two additional vignettes present a more macroscopic view of a Mosaic portfolio or enterprise in a similar time frame. These vignettes represent events that would require top-level management and priority setting across the Mosaic pipeline from R&D through fielding:

- Vignette (Portfolio) D—Managing the Pipeline I
- Vignette (Portfolio) E—Managing the Pipeline II.

During day 1 of the exercise, players explored whether the current DoD requirements, resourcing, and acquisition systems are compatible with the vignettes, and, if not, why.

Figure A.2
Mosaic Tile–Level Vignettes (Capability Threads)

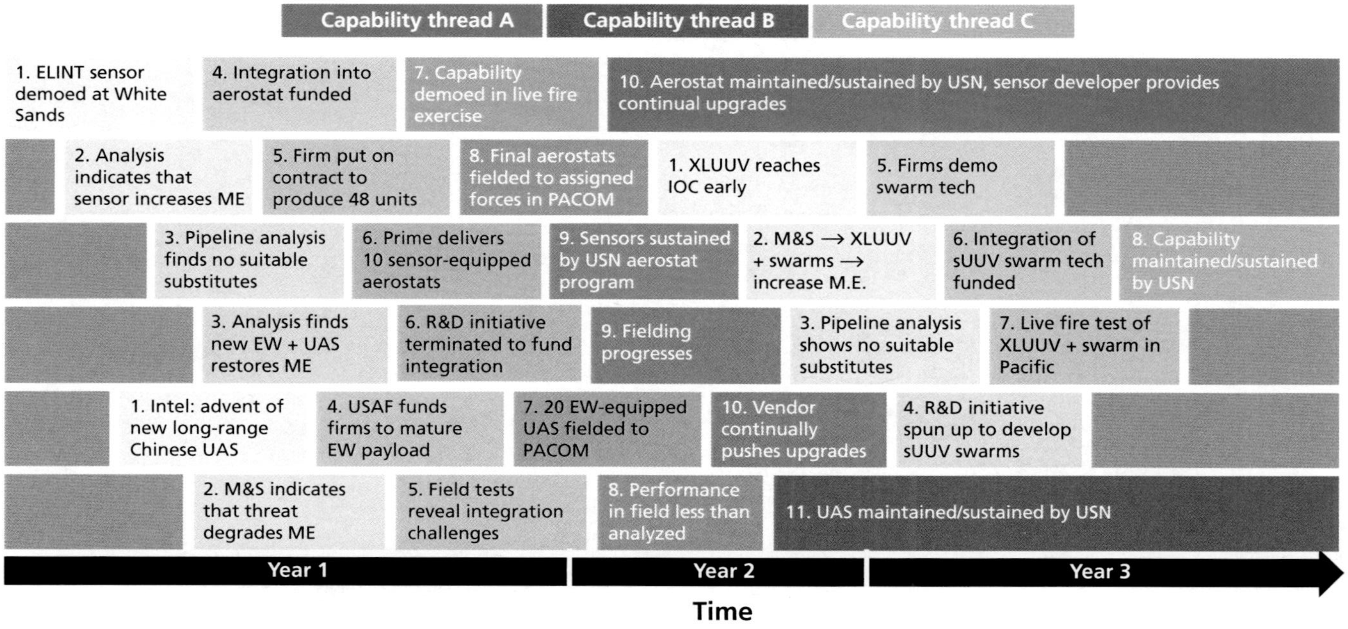

NOTES: ME = mission effectiveness; PACOM = Pacific Command; sUUV = small unmanned underwater vehicle. Placement of steps along time axis for graphical purposes only. No information is conveyed in the width or precise placement of individual steps.

Mosaic Tile–Level Vignettes (A, B, and C)

Year 2030
Vignette (Capability Thread) A—Leap Forward Sensor Tech

- A.1. A small firm demonstrates a promising new ELINT sensor on a medium rotary-wing UAS at White Sands.
- A.2. Analysis shows 20% improvement in mission (ASuW) effectiveness in 20% of scenarios analyzed with 80% probability if the sensor is integrated onto either a medium UAS or an aerostat.
- A.3. Analysis shows the sensor outperforms other fielded ELINT sensors as well as those already in the enterprise-wide development pipeline.
- A.4. An in-year decision is made to fund integration of the sensor onto a fielded aerostat.
- A.5. The small firm is put on contract to produce enough units of the sensor to achieve the net-mission effectiveness improvement as government furnished equipment (GFE) provided to the prime aerostat sustainment contractor for integration.
- A.6. The prime contractor delivers and installs an initial set of sensors onto aerostats assigned to a Naval task force ahead of a Pacific exercise, at an additional, unplanned cost representing 5% of the aerostat's program's yearly budget.
- A.7. The capability is demonstrated in live fire exercises as part of a kill chain that uses aerostats to cue land-based fires against naval SAMs.

Vignette (Capability Thread) B—Emergent Critical Requirement Gap

- B.1. Intel reports a previously unknown Chinese long-range UAS is being tested that is capable of providing near-real-time situational awareness of U.S. movements in potential future engagements.
- B.2. Analysis confirms that, unimpeded, the new UAS threat may degrade mission (ASuW) effectiveness by 20% in 30% of scenarios analyzed, representing a significant requirements gap.

- B.3. Analysis shows that fielding a previously prototyped yet never fielded Air Force–developed RF [radio frequency] effector payload to group 3 (< 1,320 MGTOW [maximum gross takeoff weight]) UASs could function as an effective countermeasure to the Chinese UAS and largely mitigate net-mission effectiveness losses. Further, there is a potential force multiplication effect provided if this RF effector is fielded in concert with the sensor in Capability Thread A. Analysis suggests that if the RF effector and ELINT sensor are both fielded there is an 70% probability that anticipated mission (ASuW) effectiveness increase of as much as 30% in 35% of future scenarios analyzed.
- B.4. The Air Force contracts with multiple companies (totaling four years, $166M) to ramp up on the mothballed payload, mature the technology, and explore its performance on 7 different existing platforms, as well as consider adapting existing platforms not currently in DoD use.
- B.5. Field tests are conducted at the Air Force Test Center in California that reveal unforeseen challenges in integrating the RF effector payload onto existing Group 3 UAS.
- B.6. A decision is made to retire an U.S. Air Force R&D project to overcome integration challenges and expedite fielding of the system.

Year 2031
Vignette (Capability Thread) A (Continued)

- A.8. The remaining aerostats are delivered with the new sensor to assigned forces in PACOM.
- A.9. The new sensors are sustained by the Navy's aerostat program. The aerostat prime contractor's budget is plussed up to handle the additional sustainment cost of the new ELINT sensors for the next 2 years.

Vignette (Capability Thread) B (Continued)

- B.7. The contractor team delivers an initial batch of fast moving and attritable Group 3 UAS equipped with new payload; they are deployed to assigned forces in PACOM.
- B.8. On deployment, performance of the payload on the UAS operating in PACOM is significantly less than analyzed, however the contractor continues to try configurations and upgrades on the digital twin residing in the CONUS [contiguous United States] test range.
- B.9. Integration and fielding progress, since analysis shows that the contribution to net-effectiveness is still positive, with no change to the probability (15%) that the effector can work in concert with the ELINT sensor to provide a significant force multiplicative benefit.
- B.10. The contractor team pushes periodic upgrades to the fielded systems, slowly increasing its effectiveness.

Vignette (Capability Thread) C—New Force Multiplier

- C.1. The Navy approves a new autonomous XLUUV for IOC [initial operating capability] earlier than expected.
- C.2. Analysis shows a net-mission effectiveness improvement of 90% in 40% of scenarios with 50% probability if the XLUUV is equipped with small, loitering UUV munitions capable of autonomously swarming.
- C.3. Analysis of the existing and pipeline commercial and government capabilities finds no promising near-ready solutions.
- C.4. DoD initiates a new requirement and begins a ($125M) R&D initiative to develop the capability.

Year 2032
Vignette (Capability Thread) A (Continued)

- A.10. The new aerostats sensors continue to be maintained and supported by U.S. Navy. The prime contractor subs out work to the initial developer to provide upgrades to the ELINT sensor.

Vignette (Capability Thread) B (Continued)

- B.11. The Group 3 UAS is maintained and supported by the U.S. Navy.

Vignette (Capability Thread) C (Continued)

- C.5. One of the firms involved in the R&D effort demonstrates a low cost sUUV munition with a limited expected lifespan but capable of being fielded rapidly. Another firm involved in the effort demonstrates AI heuristics that successfully control sUUVs in a limited scenario.
- C.6. A decision is made to integrate new algorithms on the low cost sUUV munition and integrate it to the existing XLUUV. R&D continues to search for longer term solutions. A new contract is created to transition the technology and produce low cost prototypes. This will be done in parallel with the continued XLUUV production.
- C.7. A successful live fire test of XLUUV-deployed sUUV swarm is performed in the Pacific region during joint exercises.
- C.8. The XLUUV + sUUV swarm prototype capability is sustained by the U.S. Navy.

Mosaic Portfolio–Level Vignettes (D and E)

Year 2030
Vignette (Portfolio) D—Managing the Pipeline I

Service and COCOM analysis presented during [fiscal year 2031] program review points to a need to accelerate investments in enablers—sensors, munitions, data links C2ISR [command, control, intelligence, surveillance, and reconnaissance] capabilities. To free up necessary resources, two ACAT I programs are curtailed:

- A major development competition for a manned multifunctional fighter is terminated.
- Purchase quantities for a manned surface combatant are reduced by 10% across the FYDP [Future Years Defense Program].

DoD uses these freed resources—nearly $3B in total—to expedite and expand the production of previously developed platforms that were explicitly designed to host modular payloads of various types. These platforms include:

- Three new UAS variants
- A ship-launched aerostat
- An XLUUV
- Three UUV variants
- Three USV variants
- An unmanned aerial C2 Node.

Characteristics of these programs include:

- Modal platform service life is 5 years
- Mean unit cost is $15,000,000
- Mean development-to-fielding time is 2.5 years.

DoD also funds the modernization of existing platforms to accommodate the integration of new enablers. For example, a subset of the C-130 inventory is modernized to conduct mid-air deployment and recovery of UAVs variants that could alternatively be outfitted with sensor suite, EW payloads, air-to-air or land-attack missiles.[1]

DoD also initiates a set of 30 new programs, transitioning mature technologies from the S&T base. The majority of programs are for enablers and are destined for integration onto autonomous platforms.

All of the platforms and new R&D programs meet the criteria of ACAT III programs according to existing acquisition policy. According to the DoD plan, the number of unique operationally viable ASuW kill chains available to the force in 2036 will be 10X that available to the force in 2030. Within the set of 30 new programs are the following enablers:

- A loitering UAS munition with reduced detectability
- Deployable airborne comms relays that can maintain comms for short bursts (hours) before falling from the sky (single use)
- A disaggregated mobile mine capable of separating and re-combining to avoid detection in contested environments
- Long-range Caesium magnetometer capable of detecting enemy ships at longer ranges
- A Zero-trust C2 network enclave
- Nesting torpedo capable of deploying various air and sea-borne automated munitions.

The Department exploits the prior decade's efforts to engage non-traditional suppliers, and within a year, the composition of the vendor pool has significantly shifted. Characteristics of 2030 ASuW technology vendors include:

- Median vendor has 60 employees and takes in $50 million in revenue.
- Most vendors are not publicly traded.

---

[1]   The midair deploy capability was recently demonstrated by Dynetics under the auspices of a DARPA program. The recovery capability demonstration is expected to take place soon (Dynetics, 2020).

- Modal number of programs supported by a vendor is 1. Maximum is 6.
- More than 50% of vendors are first-time DoD vendors.

The portfolio shift also involves a significant surge in reallocated spending to adapt commercial off-the-shelf (COTS) and government off-the-shelf (GOTS) technologies.

Years 2032–2033
Vignette (Portfolio) E—Managing the Pipeline II
The preceding years' investments in ASuW enablers yield a surge in capabilities on the cusp of being ready to field to forces in INDOPACOM. Examples of new capabilities on the menu for fielding include various types of autonomous aerial and maritime systems, attritable platforms, data links, C2 nodes, decision aids, sensors, munitions, small satellites, and counter-UAS systems. In addition to the new capabilities, about a dozen upgraded systems are near IOC.

To field as many of these as possible, DoD rebalances its overall portfolio away from R&D, moving $2B from prototyping and systems development to production and O&S [operations and support]. Operational test events are curtailed for the subset of capabilities that are to be integrated on previously demonstrated platforms, so long as virtual testing confirms a net-positive contribution to mission effectiveness.

By year's end, a stream of platforms and capabilities begin arriving at Guam, Kadena, Sasebo, and Subic Bay. Units begin experimenting and two large Joint, live fire exercises showcase six, previously undisclosed ASuW kill chains.

References

Alkire, Brien, Sherrill Lingel, and Lawrence M. Hanser, *A Wargame Method for Assessing Risk and Resilience of Military Command-and-Control Organizations*, Santa Monica, Calif.: RAND Corporation, TL-291-AF, 2018. As of April 26, 2021: https://www.rand.org/pubs/tools/TL291.html

Anton, Philip S., Brynn Tannehill, Jake McKeon, Benjamin Goirigolzarri, Maynard A. Holliday, Mark A. Lorell, and Obaid Younossi, *Strategies for Acquisition Agility: Approaches for Speeding Delivery of Defense Capabilities*, Santa Monica, Calif.: RAND Corporation, RR-4193-AF, 2020. As of August 27, 2020: https://www.rand.org/pubs/research_reports/RR4193.html

Arena, Mark V., Brian Persons, Irv Blickstein, Mary E. Chenoweth, Gordon T. Lee, David Luckey, and Abby Schendt, *Assessing Bid Protests of U.S. Department of Defense Procurements: Identifying Issues, Trends, and Drivers*, Santa Monica, Calif.: RAND Corporation, RR-2356-OSD, 2018. As of May 3, 2021: https://www.rand.org/pubs/research_reports/RR2356.html

Bartels, Elizabeth M., Jeffrey A. Drezner, and Joel B. Predd, *Building a Broader Evidence Base for Defense Acquisition Policymaking*, Santa Monica, Calif.: RAND Corporation, RR-A202-1, 2020. As of April 26, 2021: https://www.rand.org/pubs/research_reports/RRA202-1.html

Chairman of the Joint Chiefs of Staff Instruction 5123.01H, *Charter of the Joint Requirements Oversight Council (JROC) and Implementation of the Joint Capabilities Integration and Development System*, Washington, D.C.: Office of the Chairman of the Joint Chiefs of Staff, August 31, 2018. As of April 26, 2021: https://www.jcs.mil/Portals/36/Documents/Library/Instructions/CJCSI%205123.01H.pdf?ver=2018-10-26-163922-137

Clark, Bryan, Dan Patt, and Harrison Schramm, *Mosaic Warfare Exploiting Artificial Intelligence and Autonomous Systems to Implement Decision-Centric Operations*, Washington, D.C.: Center for Strategic Budgetary Assessments, 2020. As of December 17, 2020: https://csbaonline.org/uploads/documents/Mosaic_Warfare_Web.pdf

Colombi, John, Bryan Bentz, Ryan Recker, Brandon Lucas, and Jason Freels, "Attritable Design Trades: Reliability and Cost Implications for Unmanned Aircraft," *2017 Annual IEEE International Systems Conference (SysCon)*, IEEE, 2017.

Deptula, David A., Heather Penney, Lawrence Stutzriem, and Mark Gunzinger, *Restoring America's Military Competitiveness: Mosaic Warfare*, Arlington, Va.: Mitchell Institute for Aerospace Studies, September 2019. As of December 17, 2020:
http://docs.wixstatic.com/ugd/a2dd91_29e021b297f2492ca7f379d31466ad0c.pdf

Department of Defense 7000.14-R, *Department of Defense Financial Management Regulation*, Washington, D.C.: Office of the Under Secretary of Defense (Comptroller)/Chief Financial Officer, May 2019. As of April 27, 2021:
https://comptroller.defense.gov/fmr.aspx

Department of Defense Directive 5000.01, *The Defense Acquisition System*, Washington, D.C.: Office of the Under Secretary of Defense for Acquisition and Sustainment, September 9, 2020. As of April 26, 2021:
https://www.esd.whs.mil/Portals/54/Documents/DD/issuances/dodd/500001p.pdf?ver=2020-09-09-160307-310

Department of Defense Directive 7045.14, *The Planning, Programming, Budgeting, and Execution (PPBE) Process*, Washington, D.C.: U.S. Department of Defense, January 25, 2013, incorporating change 1, August 29, 2017. As of April 27, 2021:
https://www.esd.whs.mil/Portals/54/Documents/DD/issuances/dodd/704514p.pdf?ver=2017-08-29-132032-353

Department of Defense Instruction 5000.02, *Operation of the Adaptive Acquisition Framework*, Washington, D.C.: Office of the Under Secretary of Defense for Acquisition and Sustainment, January 23, 2020. As of April 26, 2021:
https://www.esd.whs.mil/Portals/54/Documents/DD/issuances/dodi/500002p.pdf?ver=2020-01-23-144114-093

Department of Defense Instruction 5000.02T, *Operation of the Defense Acquisition System*, Washington, D.C.: U.S. Department of Defense, January 7, 2015, incorporating change 10, December 31, 2020. As of April 27, 2021:
https://www.esd.whs.mil/Portals/54/Documents/DD/issuances/dodi/500002tp.PDF?ver=6KTtyfGjzLqbnGyWBNDnAQ%3d%3d

Dewar, James A., Carl H. Builder, William M. Hix, and Morlie H. Levin, *Assumption-Based Planning: A Planning Tool for Very Uncertain Times*, Santa Monica, Calif.: RAND Corporation, MR-114-A, 1993. As of April 26, 2021:
https://www.rand.org/pubs/monograph_reports/MR114.html

DoD—*See* U.S. Department of Defense.

DoDD—*See* Department of Defense Directive.

DoDI—*See* Department of Defense Instruction.

Dynetics, "Dynetics' X-61A Gremlins Air Vehicle Performs Its Maiden Flight," press release, January 27, 2020. As of September 18, 2020:
https://www.dynetics.com/newsroom/news/2020/
dynetics-x-61a-gremlins-air-vehicle-performs-its-maiden-flight

FAR—*See* Federal Acquisition Regulation.

Federal Acquisition Regulation, Part 10, Market Research, March 10, 2021. As of April 29, 2021:
https://www.acquisition.gov/far/part-10

Fortunato, Evan, "STITCHES: SoS Technology Integration Tool Chain for Heterogeneous Electronic Systems," briefing slides, 2016. As of December 17, 2020:
https://ndiastorage.blob.core.usgovcloudapi.net/ndia/2016/systems/
18869_Fortunato_SoSITE_STITCHES_Overview_Long_9Sep2016_.pdf

Grana, Justin, Jonathan Lamb, and Nicholas A. O'Donoughue, *The Benefits of Fractionation in Competitive Resource Allocation*, Santa Monica, Calif.: RAND Corporation, WR-1329-OSD, 2020. As of December 17, 2020:
https://www.rand.org/pubs/working_papers/WR1329.html

———, *Findings on Mosaic Warfare from a Colonel Blotto Game*, Santa Monica, Calif.: RAND Corporation, RR-4397-OSD, 2021. As of December 17, 2020:
https://www.rand.org/pubs/research_reports/RR4397.html

Landree, Eric, and Richard Silberglitt, *Application of Logic Models to Facilitate DoD Laboratory Technology Transfer*, Santa Monica, Calif.: RAND Corporation, RR-2122-OSD, 2018. As of April 27, 2021:
https://www.rand.org/pubs/research_reports/RR2122.html

Lempert, Robert J., Steven W. Popper, and Steven C. Bankes, *Shaping the Next One Hundred Years: New Methods for Quantitative, Long-Term Policy Analysis*, Santa Monica, Calif.: RAND Corporation, MR-1626-RPC, 2003. As of April 27, 2021:
https://www.rand.org/pubs/monograph_reports/MR1626.html

Lofgren, Eric, "Budget Reform," *Acquisition Talk*, blog post, undated. As of April 28, 2021:
https://acquisitiontalk.com/budget-reform/

Mayer, Lauren A., Mark V. Arena, Frank Camm, Jonathan P. Wong, Gabriel Lesnick, Sarah Soliman, Edward Fernandez, Phillip Carter, and Gordon T. Lee, *Prototyping Using Other Transactions: Case Studies for the Acquisition Community*, Santa Monica, Calif.: RAND Corporation, RR-4417-AF, 2020. As of April 27, 2021:
https://www.rand.org/pubs/research_reports/RR4417.html

McKernan, Megan, Jeffrey A. Drezner, and Jerry M. Sollinger, *Tailoring the Acquisition Process in the U.S. Department of Defense*, Santa Monica, Calif.: RAND Corporation, RR-966-OSD, 2015. As of December 17, 2020:
https://www.rand.org/pubs/research_reports/RR966.html

O'Donoughue, Nicholas A., Samantha McBirney, and Brian Persons, *Distributed Kill Chains: Drawing Insights for Mosaic Warfare from the Immune System and from the Navy*, Santa Monica, Calif.: RAND Corporation, RR-A573-1, 2021. As of April 27, 2021:
https://www.rand.org/pubs/research_reports/RRA573-1.html

Office of the Deputy Director for Engineering and Office of the Under Secretary of Defense for Research and Engineering, *Mission Engineering Guide*, Washington, D.C., November 2020. As of April 27, 2021:
https://ac.cto.mil/wp-content/uploads/2020/12/MEG-v40_20201130_shm.pdf

U.S. Code, Title 10, Armed Forces. As of April 30, 2021:
https://www.govinfo.gov/app/details/USCODE-2014-title10/

U.S. Department of Defense, *Manual for the Operation of the Joint Capabilities Integration and Development System*, J-8, August 31, 2018. As of April 27, 2021:
https://www.acq.osd.mil/jrac/docs/2018-JCIDS.pdf

Wagner, Andrew, "Creating Adaptable Technology for the Military," *Government Matters*, December 4, 2019. As of December 17, 2020:
https://govmatters.tv/creating-adaptable-technology-for-the-military/